Programming and Scheduling Techniques

Planning is an important management function and its effective execution is crucial to ensure the success of any project. This second edition of Thomas Uher's and Adam Zantis' textbook maintains its focus on operational rather than strategic aspects of programming and scheduling of projects, providing the reader with the practical planning skills needed to be successful.

Unlike most other textbooks that largely focus on the critical path method, *Programming and Scheduling Techniques* includes a comprehensive review of a range of practices used around the world. Topics covered in this thoroughly revised edition include:

- deterministic scheduling techniques including the bar chart, the critical path method, the critical chain method, the multiple activity chart and the line of balance
- a comparison of the critical path and critical chain scheduling techniques
- options for computer-based scheduling
- stochastic scheduling techniques including the critical path method based on Monte Carlo simulation and the Program Evaluation and Review Technique (PERT)
- risk in scheduling
- work study.

By covering a broad range of scheduling techniques this book is suitable for those planning projects in any industry, particularly in interdisciplinary or international contexts.

Learning activities, step-by-step guides, and a downloadable answers booklet make sure no reader is left behind. Written for students studying undergraduate and postgraduate architecture, building, construction/project management, quantity surveying, property development and civil engineering programs.

Thomas E. Uher was an Associate Professor in the Faculty of the Built Environment at the University of New South Wales between 1978 and 2009. He presently works as a consultant on project management, contract administration and partnering, and as an Adjudicator of payment claim disputes under the NSW, Queensland and Victorian Security of Payment Acts. He is author of over 100 journal articles and 3 books.

Adam S. Zantis has been working as a project planner and manager in the construction industry for the past seven years. Adam's expertise in planning and scheduling has been recognised by the University of New South Wales, where he currently lectures.

Programming and Scheduling Techniques

Second Edition

Thomas E. Uher and
Adam S. Zantis

Spon Press
an imprint of Taylor & Francis

LONDON AND NEW YORK

First edition published 2003
by UNSW Press

This edition published 2011
by Spon Press
2 Park Square, Milton Park, Abingdon, Oxon OX14 4RN

Simultaneously published in the USA and Canada
by Spon Press
711 Third Avenue, New York, NY 10017

Spon Press is an imprint of the Taylor & Francis Group, an informa business

British Library Cataloguing in Publication Data
A catalogue record for this book is available from the British Library

Library of Congress Cataloging in Publication Data
Uher, Thomas E.
Planning and scheduling techniques / Thomas E. Uher and Adam S. Zantis.
p. cm.
Includes bibliographical references and index.
1. Project management. 2. Strategic planning. 3. Production scheduling. I. Zantis, Adam S. II. Title.
T56.8.U38 2011
658.4'04–dc22
2010050588

ISBN: 978-0-415-60168-9 (hbk)
ISBN: 978-0-415-60169-6 (pbk)
ISBN: 978-0-203-83600-2 (ebk)

Typeset in Goudy
by RefineCatch Limited, Bungay, Suffolk

Printed and bound in Great Britain by
TJ International Ltd, Padstow, Cornwall

Contents

Tables

Figures

Abbreviations

ACWP	Actual cost for work performed
AREA	Area/department
AS	Activity slack
BAC	Budget at completion
BCWP	Budget cost for work performed
BCWS	Budget cost for work schedule
CC	Critical chain
CCM	Critical chain management
CCS	Critical chain scheduling
CIM	Control interval and memory
CONS	Construction department
CONT	Contracts department
CPI	Cost performance index
CPM	Critical path method
CV	Cost variance
DEP	Department
EAC	Estimate at completion
EFD	Earliest finish date
EPS	Enterprise project structure
ES	Event slack
ESD	Earliest start date
ESTM	Estimating department
EV	Earned value
FF	Free float
FINC	Financial department
FTS	Finish-to-start
FTF	Finish-to-finish
HRMG	Human resource management department
ID	Identification code
ITEM	Item name
LSD	Latest start date
LFD	Latest finish date

LOB	Line of balance
LOCN	Location
LSM	Linear scheduling method
LSMh	Linear scheduling model
MAC	Multiple activity chart
MAX	Maximum
MBO	Management by objectives
MILE	Milestone
MIN	Minimum
OBS	Oraganisational breakdown structure
PC	Percentage complete
PERT	Program evaluation and review technique
PRCH	Purchasing department
RESP	Responsibility
RPM	Repetitive project modelling
RUF	Resource utilisation factor
SPI	Scheduled performance index
STS	Start-to-start
STF	Start-to-finish
SV	Scheduled variance
TF	Total float
TLS	Tender letting schedule
TOC	Theory of constraints
VPM	Vertical production method
WBS	Work breakdown structure

Preface

Planning is the first functional step in any production process. It requires that every task essential in developing a construction project is identified, and carefully incorporated and integrated into an overall development programme that ensures successful project outcomes. Without an effective plan, there can be no control over the production process.

In any organisation, planning is essential at all management levels. Top managers are concerned with strategic and business planning, while middle and lower managers develop operational plans which formulate specific operational strategies for achieving the objectives set out in the strategic and business plans. This book addresses operational rather than strategic aspects of planning of construction projects. It describes specific scheduling techniques and processes commonly used in the construction industry. While used mainly at the construction stage, the described techniques and processes are suitable for application across all the stages of the project lifecycle.

Many books have been written on aspects of construction scheduling, but they largely focus on the critical path method and do not provide a comprehensive review of a range of scheduling techniques. This book attempts to redress this problem.

While this book serves as a reference for construction industry practitioners, it has mainly been written as a text and reference material for students studying architecture, building, construction management, civil engineering, and for quantity surveying undergraduate and postgraduate programmes.

The second edition of this book benefits from having Adam Zantis as second author. Adam has injected into this new edition his practical knowledge of applying various scheduling techniques in the construction industry. This is reflected in inclusion of status reporting in Chapter 6, *Project control*, the substantial revision of the material in Chapter 7, *Critical path scheduling by computer*, a revision of the material in Chapter 8, *Critical chain scheduling*, and inclusion of the new material on productivity rate databases in Chapter 11, *Work study*, and on risk contingency calculations in Chapter 12, *Risk and scheduling*.

The format of the book remains much the same with the previously discussed scheduling techniques retained. However, the content of some chapters has been either revised or updated and, where appropriate, expanded. Scheduling techniques described in the book include the bar chart, critical path method, critical chain scheduling, multiple activity chart and line of balance. Although not strictly a scheduling technique, work study has nevertheless been retained since it assists planners and project managers in developing efficient production methods.

T. E. Uher
A. S. Zantis

Chapter 1

The concept of planning and control

1.1 Introduction

The purpose of this chapter is to introduce the concept of planning, and in particular operational planning. A systematic approach to planning will be discussed first, followed by a brief review of different types of planning activities such as strategic, operational and coordinative. A range of planning tools and techniques will then be examined, followed by a discussion on important issues relevant to planning of construction projects. In particular, the distinction will be made between time and resource scheduling. In the next section, an overview of specific planning tasks employed in individual stages of the project lifecycle will be given. Finally, examples of plans, programmes and schedules used in the construction industry will be illustrated.

Planning is one of the four main functions of management. Together with organising, control and leading, it forms the foundation pillars of effective management (Robbins and Coulter 2009). In simple terms, planning is a process of forecasting future outcomes that may be uncertain or even unknown. It means assessing the future and making provision for it by gathering facts and opinions in order to formulate an appropriate course of action. Planning thus develops a strategy and defines expected outcomes (objectives) for undertaking a specific task before committing to such a task.

Once a planning strategy has been determined and objectives defined for a specific task, the manager will select and allocate necessary resources for carrying out the work. This is referred to as 'organising'. It is the second of the four most important management functions.

Because a plan is only a forecast of some specific future events whose outcomes are uncertain, it would be unreasonable to expect it to be accurate. Realistically, the prudent manager will expect the actual progress to deviate from the plan. Accepting that some deviation will occur, the manager will look for it by regularly monitoring the progress, evaluating uncovered deviations from the plan and replanning the work accordingly. This process is referred to as 'control'. It is the third of the four important management functions.

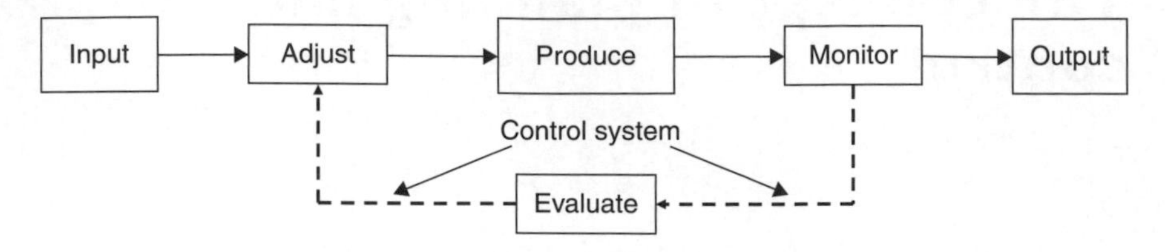

Figure 1.1 A typical production process.

'Leading' is the fourth management function. While vitally important for achieving project success, it lies outside the scope of this book. The reader is nevertheless encouraged to develop an understanding of the topic of 'leading' by reference to any of many published management textbooks.

Planning, organising and control functions are closely linked within a typical production process. This is illustrated graphically in Figure 1.1.

Planning is the foundation stone of control. It would be pointless to develop a plan if there was no attempt to control its implementation. Effective control of the production process involves its regular monitoring, evaluation and adjustment. The control process needs to be dynamic to reflect changing circumstances caused by issues such as:

- Fluctuations in the level of demand and sales
- Availability of resources
- Changes in the level of economic activity
- Changing strategies of competitors.

1.2 Planning process

If the main focus were to generate profit and increase it annually, the organisation would regard profitability as its objective and would develop appropriate business strategies for achieving it. Construction projects are no different. They are expected to be completed on schedule, within the cost budget and to the required quality and safety standards. These are the most common objectives of projects.

Establishing objectives is the first step in a typical planning process. Other steps are (Robbins and Coulter 2009):

- Forecasting
- Examining resources
- Establishing policies
- Developing alternatives
- Creating procedures and rules
- Establishing budgets

- Establishing timetables
- Determining standards.

These planning steps will now be briefly discussed.

1.2.1 Setting objectives

Planning begins by setting objectives and defining strategies for achieving them. It occurs at each organisational level and leads to the development of a comprehensive hierarchy of strategies and actions relevant to the whole organisation as well as to its individual levels of management.

Planning across the entire organisation is a complex, systematic process that requires careful coordination and integration of a wide range of activities. At the top are the objectives of the organisation as a whole. These are broken up from the top down into a set of objectives relevant for each level of organisational activity.

The overall organisational objectives may often be vague. Becoming the market leader, increasing profitability, providing the best customer service or promoting best practice serve as examples of organisational objectives that illustrate the degree of their generality.

As organisational objectives are passed down the line, they usually become more specific. For example, objectives of the construction department of a building contractor might be to improve the tender success rate, deliver construction projects ahead of time or improve people's skills. At a project level, objectives become quite specific. For example, in the project period they define its cost and the required quality standards. They may also define safety performance and the maximum permissible level of contractual and industrial disputes. With such specific objectives in place, the project manager is able to assess the actual level of progress and performance once the project is under way.

In setting objectives, particularly those at the organisational level, the company's management should then take a broader view of objectives by asking themselves what business they are in and what their customers really pay them for. If, for example, Hewlett-Packard, which is better known for producing office machines, had not defined its business as 'supplying information', its growth could have been inhibited in the fast-expanding field of information technology.

Similarly, in setting their own objectives, the subordinate managers need to identify what the organisation really wants from them and from their functional units or projects. This process will ensure that objectives are correctly defined and properly integrated cross the entire organisation. They will then form an integrated network or 'a means–ends chain' (Robbins and Coulter 2009) where the objectives at a lower level (referred to as means) need to be satisfied if the objectives at the next higher level (referred to as ends) are to be met.

Setting objectives from the top down has its advantage in that subordinate managers know what they need to achieve within their sphere of responsibility.

Sometimes, however, managers down the line may become frustrated when their input is not sought from above in setting the objectives. This may impact negatively on managers' motivation and lead to inefficiency.

The ideal scenario is when managers down the line are asked to contribute to setting objectives and the formulation of plans for reaching them. For example, state managers of a construction company may be required to forecast the future volume of work in their respective states and suggest strategies for increasing turnover. This information would help top management in developing the overall corporate objectives. When objectives are set jointly by subordinates and their superiors across different levels of organisation, and rewards allocated on the basis of achieving progress, management practice is referred to as 'management by objectives' (MBO). It is a popular and well-established system for setting objectives and ensuring their successful accomplishment (Robbins and Coulter 2009).

1.2.2 Forecasting

Forecasting is the estimation of future controllable and uncontrollable events and opportunities pertinent to an organisation's business activities. It involves systematic assessment of future conditions, such as economic climate, political and social issues, future demand for goods and services, changes in the population growth, and the like.

Forecasts are by their very nature always wrong. It is therefore the manager's task to continually revise and update the information upon which the forecasts and ultimately decisions are made.

1.2.3 Examining resources

The achievement of objectives is largely dependent on the correct allocation of resources in the form of people, materials, plant, equipment, money and time, and their effective management.

Even well-defined plans may go astray. When, for example, a construction project falls behind its schedule and the project manager is unable to speed it up by simply making the committed resources work harder or by improving the work method, often the only remedy available is to inject additional resources. These, however, may not be available when needed, or may not have the required skill or capacity. Furthermore, the cost of injecting additional resources may be prohibitive, particularly in the case of heavy plant or equipment. The manager's task is to allocate enough resources to achieve the given objectives and to make contingency plans for securing additional resources for times when they might be needed.

1.2.4 Establishing policies

Policies are guides for thinking. They govern the execution of activities within an organisation. They are the foundation of an organisation and provide a broad pathway for workers to follow in order to achieve an objective. The recruitment and the promotion policies serve as examples of typical organisation policies.

Organisations form their own policies, but sometimes policies are imposed by external sources such as governments. The New South Wales Government procurement policy (NSW Government 2004) is an example of a policy externally imposed on organisations bidding for government work.

1.2.5 Developing alternatives

No planning task is accomplished without the development of plausible alternatives. All possible courses of action need to be identified and evaluated, even those that may appear to be unusual or even ridiculed by some. It may well be that the most unusual alternative will provide the best solution for achieving the objectives if assessed objectively.

1.2.6 Creating procedures and rules

Within the bounds of the organisation's policies, procedures and rules provide a precise recipe for the manager to follow in taking action. Procedures define a logical sequence of steps to be followed in decision-making. Rules are agreed to in advance. They define specific conditions of various procedural steps. For example, a contractor has a policy for awarding work to subcontractors on the basis of competitive tendering. The tendering procedure then defines, step by step, the contractor's approach to administering competitive tendering to subcontractors. One of the rules defined in the tendering procedure may be that tenders must be sought from at least three subcontractors per trade contract.

1.2.7 Establishing budgets

The purpose of budgets is to work out beforehand an expected outlay of resources necessary for the accomplishment of the objectives. Depending on the planning task, budgets may be expressed in financial terms, which is the most common form, or in some other terms, such as duration, labour hours or the quantity of materials needed.

Budgeting is both a planning and a control task. The accurate knowledge of where the resources have actually gone to is essential in measuring efficiency in the production process. It also assists in future planning.

1.2.8 Establishing timetables

Like cost and other resources, time is an important element in planning. Practically all planning tasks are constrained by time. There is almost always a deadline for accomplishing the work.

In the absence of any time constraints, a manager would be able to establish the completion date of a project from the planned quantity of work and the required level of resources employed to carry out the work. However, most construction projects are rigidly constrained by time. Delays in completion have serious financial implications for the parties concerned. The development of a project schedule often requires the project manager to plan from the end, that is, from the deadline set for the completion of the project. In such situations, the planning task is focused more on fitting all of the required work and the necessary resources into a rigid time-frame, which often results in inefficiencies and high cost, rather than on developing a plan optimised in terms of time, cost and other resources.

1.2.9 Determining standards

A plan is an essential element of control. As such it is required to stipulate expected performance standards in terms of time, cost, the use of resources and the like. By assessing progress and performance of a project in the production process the manager is then able to establish whether the expected performance standards are being met.

1.2.10 Reviewing

Before a plan for a particular task is finalised and implemented, the manager will review the plan and critically reassess the overall planning strategy, particularly sequences, methods, resources and budgets. At this point in time, an understanding of the time component of the planning strategy would have been developed and other alternative planning strategies may have been tested by the production of alternative timetables.

1.2.11 Implementing

No matter how good the plan is, its effectiveness will ultimately depend on the degree of commitment the organisation's personnel give it. The starting point in developing the required degree of commitment is to involve those people who are going to be affected by the plan in its development. If the plan requires the adoption of new processes or techniques, people involved in such processes may need to improve their existing skills or acquire new ones through training. Effective communication at this point in time will help ensure that the project plan is implicitly and explicitly accepted of by the project team.

1.3 Types of planning activities

Planning occurs at all levels of an organisation, but the emphases vary from level to level. At the highest level is the 'strategic planning', which determines an overall business strategy such as mission and objectives. At lower levels, 'operational' and 'coordinative' planning activities take place.

1.3.1 Strategic planning

Strategic planning is defined by Bryson and Alston (2004: 3) as 'a disciplined effort to produce fundamental decisions and actions that shape and guide what an organisation is, what it does, and why it does it'. While the definition suggests that an organisation is an ongoing corporate entity, the concept of strategic planning is equally relevant to an organisation of a limited life span, such as a construction project organisation.

Strategic plans are long-term, covering a period of between three to five years. They include the formulation of a statement of mission, objectives, and broader strategies for accomplishing the stated objectives.

1.3.2 Operational planning

In comparison to strategic plans, operational plans are medium- to short-term with a time-scale defined in days, weeks or months. The main purpose of operational plans is to develop detailed strategies for achieving specific objectives. Construction schedules are an example of operational plans.

1.3.3 Coordinative planning

Coordinative planning establishes policies, procedures, rules, tactics and strategies through which strategic and operational planning are linked together. A risk management plan or a quality management plan are examples of coordinative planning.

1.4 Planning tools and techniques

Numerous planning tools and techniques have been developed to assist managers with planning. A comprehensive description of planning tools and techniques can be found in Robbins and Coulter (2009). Some of those will now be briefly reviewed.

1.4.1 Environmental scanning

Environmental scanning is a systematic screening of information from which the manager is able to detect emerging trends. In some industries, such as IT, advances in technology are rapid. Managers need to be constantly on the alert for new advances and trends that could maintain or improve their organisation's competitiveness. The most commonly used techniques are competitor intelligence, global scanning and scenario-building.

1.4.2 Forecasting

Forecasting assists the manager to predict future outcomes. This topic has already been briefly discussed in 1.2.2 above.

1.4.3 Benchmarking

'Benchmarking is the comparison of practices either between different departments within the company, or with other companies in the same industry, or finally with other industries' (McGeorge and Palmer 2002: 81). The aim of benchmarking is to search for the best practice in order to achieve superiority among competitors. It has become a popular and a widespread technique across many industries, including construction.

1.4.4 Budgeting

Budgets were defined in 1.2.7 above as plans for allocating resources to specific activities that would be performed by an organisation in its attempt to attain objectives. Budgets are formed for many different items or areas of activities, examples of which are revenue, expense or capital expenditure budgets.

1.4.5 Operational planning tools

Operational planning tools are used by managers in day-to-day problem-solving. This may involve scheduling of work using bar charts or the critical path method, allocating and levelling resources using linear programming, predicting profitability through, say, break-even analysis, assessing cost performance by way of earned value, and assessing risk in decision-making through the concept of risk management. There are many other techniques including queuing theory, marginal analysis and simulation that assist managers in decision-making.

This is the area of planning that this book addresses. The following chapters will discuss several operational planning tools used in planning and controlling construction projects.

1.5 Planning of construction projects

The foregoing discussion has addressed the general principles of planning that can be applied in any organisation. Since the main purpose of this book is to examine the scheduling of construction projects, the focus of discussion will now shift to important issues pertinent to the planning of construction projects.

A project can be said to be a particular and unique form of organisation with a limited life span. It exists for a finite period only. When its objectives have been achieved, it comes to an end. An important characteristic of projects is that they are driven by objectives that are defined at the start. The project manager assumes overall responsibility for the project and leads the project team.

Between their beginning and their ending, projects pass through a series of stages commonly known as the 'project lifecycle', through which a project is conceived, designed, constructed and commissioned. Planning activities that take place in individual stages of the project lifecycle will be discussed later in this chapter.

1.5.1 Definition of planning terms

Although the terms 'planning', 'programming' and 'scheduling' are generally used indiscriminately, they are fundamentally different from each other. Using a top-down hierarchical approach, 'planning' sits at the top. It is the overall approach to predicting a future course of action. 'Programming' is placed in the middle. It is the name given to the task of identifying activities, establishing relationships, and developing logical sequences among such activities that would depict their order of execution. 'Scheduling' sits at the bottom. It is a process of quantifying the programme. This involves, for example determining times and costs of activities, and the efficiency of allocated resources.

For the sake of simplicity and clarity, no distinction will be made between the terms 'programming' and 'scheduling' in this book. But a distinction will be maintained between 'planning' and 'scheduling'.

'Construction planning' may refer to a range of tasks concerned with determining the manner in which a job is to be carried out: budgeting, forecasting, preparing feasibility studies, or creating construction schedules. 'Construction scheduling', however, is concerned only with sequencing and timing of activities in a particular production process.

1.5.2 Hierarchy of plans and schedules

A well-planned construction project will use a series of plans and schedules. There will be those that show the planned use of time such as bar charts or critical path schedules. There will also be those that show planned use of other important resources such as labour, materials, plant and money. These will be prepared in the form of schedules, charts, graphs or histograms, with close attention paid to their coordination and integration.

While in theory it is possible to prepare a schedule for a large project that would show activities across all its lifecycle stages in great detail (for example charting each day), in practical terms such a schedule would be too large, too complex, unworkable and useless in communicating the plan to others. There is ample anecdotal evidence of schedules that cover the entire wall space of site offices. Such schedules are better suited for interior decoration of offices than for managing construction projects.

Clearly, it is necessary right from the beginning to agree on a hierarchical system of schedules for a given project. This is achieved using a 'work breakdown structure', or WBS.

The WBS assists in breaking down the total scope of the project into subsystems, elements and activities related to its design and construction. Typically, the project is divided hierarchically into five or six levels to show the required degree of detail and to ensure that every important aspect of the project has been identified. A typical example of a WBS is given in Figure 1.2, where 'Subproject A', for example, may involve the development of a brief, 'Subproject B' refers to the process of design and documentation, and 'Subproject C' represents construction of the project. Lower levels of WBS define project tasks in more detail.

Once the WBS has been defined, scheduling of the work begins for each level. The format of schedules developed at each level of the WBS is likely to

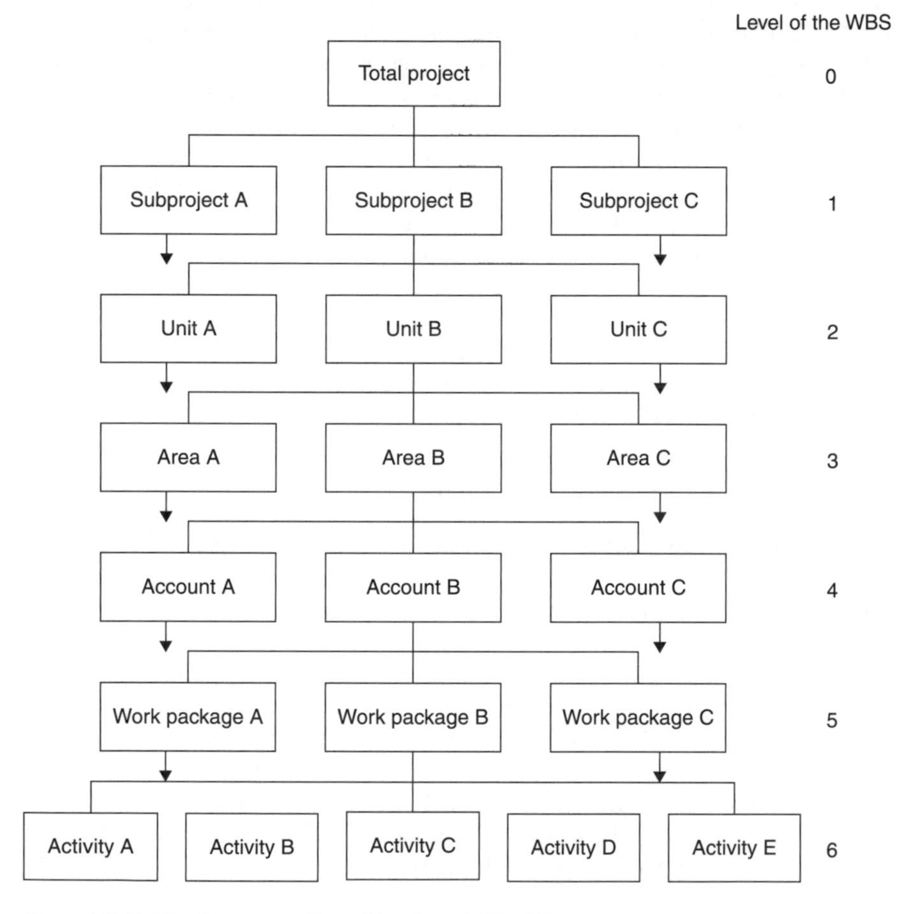

Figure 1.2 WBS of a project (from Hamilton 1997: 86).

vary, as will the amount of detail shown. Those prepared to higher levels of the WBS will show a graphic overview of the total project period with only major work sequences, while the amount of detail will increase for lower-level schedules.

The programme hierarchy becomes the basis for planning work activities, monitoring and evaluating their progress and exercising control. At the top is a long-term schedule for the entire project. Its time-scale will commonly be in weeks or months and will show the overall strategy for the job. Specifically, it will show the start and completion of each project stage, major activities and their interdependencies, the key resources needed and the provision for delay contingency. It also shows the key target dates that must be met across the entire project lifecycle. Depending on the size of the project and its overall period, it may well be that more than one long-term schedule will be prepared. For example, for the project in Figure 1.2 the total of four long-term schedules may be prepared, one for level 0 and three for level 1 of the WBS.

A 'medium-term' schedule is prepared next. It commonly shows a period of 8–12 weeks of work within the overall long-term schedule. It is more detailed than a long-term schedule and commonly shows work tasks on a weekly scale. For the project in Figure 1.2, three medium-term schedules would be prepared for level 2 of the WBS.

A 'short-term' schedule shows the greatest level of detail, such as resources. The main activities are broken down into subactivities in order to determine the extent of resource requirements. A short-term schedule commonly shows between one- and four-week production periods and its scale is in days. Because the main resources have already been committed, the task is then to allocate the work to the committed resources. Short-term schedules are thus said to be resource-driven.

A short-term schedule may further be broken down into a series of 'daily schedules' for specific tasks. A daily schedule would simply list activities to be performed in one day together with their durations expressed in hours. Alternatively, a simple bar chart can be used for this purpose.

An overall long-term schedule is a reference point for schedules on the next lower level of WBS, which in turn becomes a reference point for schedules on levels further down. Preparing various WBS schedules manually would be difficult if not impossible given the need to closely integrate and coordinate a very large number of activities, and in consideration of the fact that a change to a higher level schedule would require lower level schedules to be amended accordingly. This problem is overcome by the application of CPM (critical path method) software, such as Primavera P6, where schedules at various levels of WBS are linked together using a system of codes. A change to a higher level schedule is automatically promulgated to all the other schedules within WBS.

1.5.3 The concept of scheduling

Scheduling involves answering the questions, such as 'When can the work be carried out?' 'How long will it take?' and 'What level of resources will be needed?'

Scheduling is concerned with sequencing and timing. Since time is money, it is also concerned with cost.

Scheduling is performed using appropriate operational planning tools such as a bar chart or the CPM. Scheduling is a modelling task that assists in developing a desired solution for a problem and tests its validity. Through modelling, the physical size of a problem (say the development of an overall construction schedule for a large construction project) is scaled down to a smaller number of representative activities that are then scheduled by an appropriate scheduling tool. The solution is checked and validated for accuracy. The derived schedule is used to plan, organise and control construction of the project.

To check time progress, individual parts of the job are tracked to determine whether they are being completed within the specified time limits. Such information is then analysed to determine where the problems are.

The most common form of a model in scheduling is a graphic chart or network generated either by hand or by computer. The form and shape of a chart varies from technique to technique. For example, a sheet of paper lined with columns and rows represents a bar chart format, while charts generated by the CPM are rather more complex and are commonly referred to as networks (see for example Figures 3.3 and 3.4 in Chapter 3).

1.5.4 Types of schedule

This book makes reference to three types of schedule:

- A time schedule
- A resource schedule
- A target schedule.

The fundamental difference between time and resource schedules has previously been alluded to. This issue will now be discussed in more detail.

Time schedule

Sometimes a schedule may be prepared to show a logical sequence of activities with only notional information about duration of activities. The main aim would be to see the logic of the production process and its approximate duration. Schedules produced for this purpose are referred to as 'time schedules'.

A time schedule is prepared on the assumption that its activities will be given all the required resources when needed. In other words, time scheduling assumes that resources are unlimited and available when needed. This is an unrealistic assumption, since resources may simply be unavailable when needed, or available in limited quantity, size and type or technical specification. Furthermore, the assumption of unlimited resources will lead to inefficient

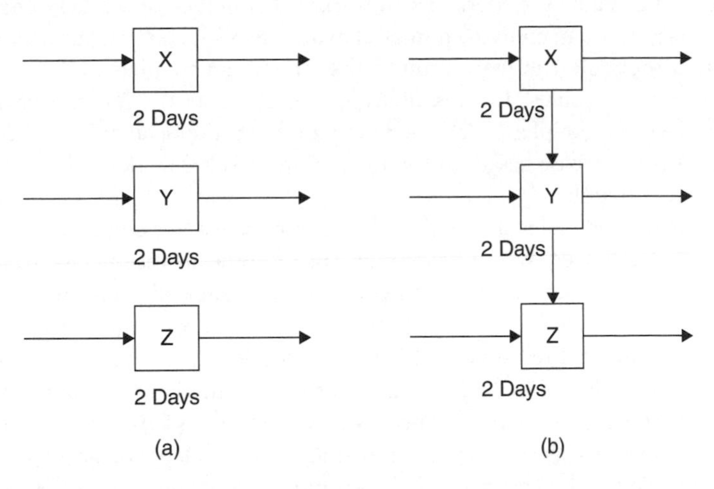

Figure 1.3 Example of time and resource schedules.

allocation of resources and the likelihood of higher cost. This is illustrated in the following example.

Let's focus on just three concurrent activities, X, Y and Z, in a schedule in Figure 1.3(a). Assume that the three activities X, Y and Z take two days each to complete. Let's assume further that each of the three activities requires access to a crane for two days. Under the assumption of unlimited resources, a project manager could order three cranes, one for each activity. The three concurrent activities X, Y and Z would then be completed in two days.

Time schedules are useful in developing an overall strategy within a broadly defined time-frame. But since they assume that resources are readily available when needed, they may not create a realistic plan of the actual production process. More realistic planning can be achieved using resource schedules.

Resource schedule

When resources are limited in availability, technical specification, cost or some other means, those resources that are actually available then drive the scheduling task. For example, a very large high-rise project may require a crew of 2,000 workers to be employed at the peak of activities. With such a large number of people working on the site, part of the project manager's task will be to ensure that the site provides the required volume of site amenities, the necessary safety equipment for the workers, and that the workers can be moved efficiently through the structure to their work stations. It may well be that the site is too confined to provide the necessary volume of amenities for so many workers. The site may also be too confined to accommodate a sufficient number of personnel hoists that would ensure speedy distribution of workers to their stations. The project manager may respond by limiting the number of workers engaged on the site (say 1,200 maximum) to match available resources. The project manager will then schedule the work around the maximum number of 1,200 workers. This approach ensures that resources are used efficiently, but the project may take longer to complete. A schedule based on the available or committed resources (in this case based on the maximum of 1,200 workers) is referred to as a 'resource schedule'.

In resource schedules the work to be accomplished is assigned to available or committed resources. When the volume of resources is insufficient to carry out the work that has been scheduled, the project manager would need to either inject more resources or reschedule the work to free over-committed resources. Injecting additional resources is likely to incur extra cost while keeping to the same schedule. Rescheduling the work around committed resources will most likely extend the project period and possibly even its cost. In organising resources in the planning stage, the project manager has the opportunity to seek an optimum relationship between the cost and time of the project. But once the project is under way and resources are committed, it is not possible for the project manager to keep the cost and time at optimum when the committed resources are insufficient to carry out the scheduled volume of work. The project manager then has either to inject more resources or to reschedule the work. Neither scenario is desirable since this is likely to increase the project cost and extend its duration.

Let's go back to the problem in Figure 1.3 and assume that only one crane is available for the scheduled work involving activities X, Y and Z. With only one crane available, it is clear that these activities cannot be performed concurrently as scheduled. The project manager would need to allocate the available crane to one activity at a time in some order of priority. Let's assume that the order is X, Y and Z. The completion of all three activities, X, Y and Z, will now take six days as illustrated in Figure 1.3(b). In both examples (a) and (b), the total volume of work will be the same: six crane-days. But the completion times and costs will be different. The difference in completion times is clearly apparent, but why would the costs be different? Since time is money, the former case would incur lower overhead costs due to a shorter schedule, but the cost of assembling and dismantling the three cranes would be higher than in the latter case.

In summary, the concept of time scheduling requires the project manager to vary the volume of resources to meet work demands. This means bringing in additional resources such as plant/equipment and people to satisfy short-term peaks in resource requirements, and removing them when they are no longer required. Apart from the extra cost associated with bringing resources in and taking them away, there is an additional administrative cost associated with planning, organising and controlling such activities.

An important observation here is that time schedules are likely to provide an overly optimistic assessment of the project period. For example, a contractor who has won a tender on the basis of a time schedule runs a serious risk of either delaying the contract due to having insufficient resources at certain times, or spending money on additional resources that must be injected to keep the project on schedule, or both.

Clearly, better control of time and cost are more likely to be achieved using resource-based scheduling. The process of managing resources, particularly in terms of allocation and levelling, will be discussed in detail in Chapter 4.

Target schedule

The term 'target date' implies that a specific activity or task must be accomplished by that date. Adding specific targets, such as starting or finishing dates, or dates at which Stages 1, 2 and 3 of the project are to be accomplished, to activities in a resource schedule results in the creation of a 'target schedule'. Target dates are commonly imposed by a contract. A schedule that is resource-based and contains target dates is a realistic scheduling tool.

1.6 Planning tasks at different stages of the project lifecycle

A project success is dependent on the effective management of each stage of the project lifecycle and the project as a whole.

The conceptual stage defines the project's scope and includes the client's needs, the main objectives and the preferred development strategy. Upon these a project brief is framed from which the design team, led by the design manager (often the project architect), designs and documents the project in the design stage. The design documentation is then used in the tendering stage to assist in selecting a general contractor. In the pre-construction stage, a successful contractor develops a strategy for building the project and then constructs it in accordance with the contract documentation in the construction stage. At the end of the construction stage, the client takes possession of the completed project. This stage is commonly referred to as commissioning. The individual stages of the project lifecycle are illustrated in Figure 1.4.

Of particular importance are the first two stages, which are concerned with defining the project concept and developing its design. This is because the capacity to control project objectives effectively (especially in terms of cost and time) diminishes as the project progresses through its lifecycle. For example, if the client alters the scope of the project at the conceptual stage, the impact on the overall project cost/time is likely to be fairly small since no design or construction has yet begun. But if the changes to the scope occur in the construction stage, the cost/time impact will be considerably greater.

Conversely, effective coordination and management of the conceptual and design stages provide an opportunity for better control of cost and time later in the construction stage. The importance of the conceptual and design stages is shown graphically in Figure 1.4.

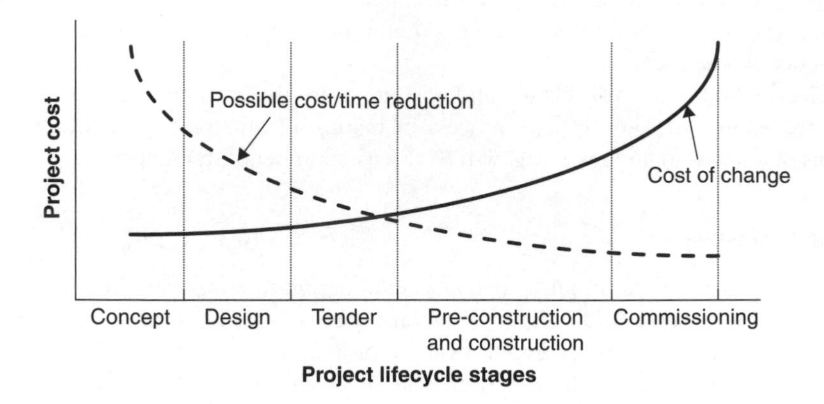

Figure 1.4 Cost of change and possible cost/time reduction across the project lifecycle.

The importance of effective lifecycle management is obvious. Since planning is one of the essential elements of effective management, it is worth examining specific planning tasks that take place in individual stages of the project lifecycle.

1.6.1 Planning at the conceptual stage

The conceptual stage is characterised by the definition of the project's scope, the development of the preferred development/construction strategy, budgeting and the formulation of a brief.

A project manager will be engaged in two distinct planning tasks: (i) strategic planning, and (ii) scheduling.

The function of strategic planning was discussed in section 1.3.1. Let's now examine the function of scheduling at the conceptual stage. Once the overall project development/construction strategy has been determined, the project manager starts work on developing an overall project schedule that shows important tasks or activities across all the stages of the lifecycle. The schedule will be a time schedule that will not show a great deal of detail but will show a logical sequence of important activities to be accomplished at each stage. It will also show relationships among activities within and across individual lifecycle stages from which the project manager will attempt to foresee potential coordination and integration problems. The degree of detail shown will be governed by the scale of the schedule, which is likely be in months or perhaps even weeks. Of particular importance is the identification of dates for major decisions.

Apart from the overall project schedule, the project manager also develops a detailed schedule for managing the conceptual stage. It will be a medium-term schedule with a time-scale in weeks. It will show a sequence of activities to be undertaken at the conceptual stage, together with the dates for the key decisions and targets. The most important target date will be that for the completion of a brief. The execution of the work at the conceptual stage requires commitment of human resources by specialist consultants. These need to be carefully assessed and built into the schedule. Due to the lack of detailed information available at this stage of the project lifecycle, the project manager needs to exercise caution when distributing final detailed schedules containing many assumptions about the information that is not available.

1.6.2 Planning at the design stage

The aim of the design stage is to design the project and prepare the necessary design and tender documentation. Design is a creative task that is often difficult to fit into a rigid time-frame. It is a time-consuming process that adds substantially to the total development period. It is also a process that is often subjected to a wide range of risks. Nevertheless, every effort is required to (i) set aside a sufficient time for design and documentation, (ii) allocate the necessary resources to ensure that the work can proceed as planned, and (iii) monitor and control the process to ensure that the work is completed on time and to a cost budget.

Planning tasks in the design stage comprise the development of:

- A design management plan
- A medium-term schedule of design activities
- A short-term schedule of weekly design tasks
- A schedule of drawings.

A design management plan is basically a strategic plan formulated for the design stage. It states the main objectives, describes strategies for achieving them, and gives budgets. For more information on strategic planning, refer to section 1.3.1.

Apart from determining the overall strategy, the project manager, together with a project team, prepares a design schedule (a medium-term schedule) showing a sequence of design and documentation activities. This schedule will need to be developed around available resources such as designers, draftspeople and computer-aided graphics equipment, and must take into account specific target dates such as dates for submission of design documentation to local councils for approval and the date for tendering. Since the design work may involve a number of separate design organisations, a design schedule is vitally important, not just for the planning of design activities but also for their coordination and integration.

A weekly design schedule will be prepared to show in detail activities and resources needed to accomplish the design work. This schedule will be used to control the everyday design production process.

Another important schedule that will be prepared in the design stage is a schedule of drawings. Since the design process brings together many different design organisations that may produce hundreds of drawings and details, a schedule of drawings will assist in monitoring the production of individual drawings. It will ensure that drawings are produced when required and distributed to the right parties.

1.6.3 Planning in the tendering stage

In the tendering stage, bidding general contractors prepare tender proposals based on tender documentation and other relevant information from which the client selects the winning tenderer.

Tendering is a form of competition for work among bidding contractors. Each bidding contractor estimates the cost of construction based on a preferred construction strategy defined in a tender schedule. In developing an appropriate construction strategy, the contractor would need to focus on issues such as:

- The type and nature of project to be built
- The site location
- Site conditions
- Contract conditions
- Alternative construction strategies
- An appropriate form of WBS.

A tender schedule is produced in the form of an overall construction programme. It is resource-based and is sufficiently detailed to enable the client to establish how the contractor will meet the project objectives.

1.6.4 Planning in the pre-construction and construction stages

After the contract has been awarded to the general contractor, the contractor begins work on developing an overall construction schedule. At first, the contractor reviews a tender schedule and highlights those tasks that may need to be modified as more up-to-date information becomes available, particularly with regard to the design. The contractor also reviews the previously defined WBS and affirms its final structure.

The WBS determines the extent of the contractor's scheduling in the pre-construction stage. Large projects constructed over a number of years will require the development of a hierarchy of schedules from the top down, starting with an overall construction schedule that will be supplemented with schedules for each level of the WBS. Smaller projects may require the development of only one overall construction schedule.

In developing a construction schedule, the contractor would commonly consider a range of issues including:

- Off-site activities, for example prefabrication
- Incomplete or missing segments of the design and documentation

- Production of shop drawings
- Lead-times and processes for approvals by authorities, consultants and the client
- Lead-times for orders of materials and equipment, and their deliveries to the site
- Off-site work of specialist contractors, particularly in the area of building services, such as air conditioning, lifts, hydraulics and electrical
- Risks associated with on-site and off-site activities that are generally outside the contractor's control
- Lead-times for delivery and installation of temporary equipment such as cranes, formwork systems and the like.

Since most construction activities are performed by specialist subcontractors, it is essential for the contractor to seek their input into scheduling. This is particularly important for coordinating and integrating the work of subcontractors and in defining the required level of resources, although this may be difficult at times due to the subcontractors' not yet being contracted to carry out work. Similarly, the contractor would need to seek input into scheduling from suppliers of materials and plant/equipment, particularly those whose delivery times are likely to require long lead-times. For example, in order to ensure timely delivery of air conditioning plant manufactured overseas, the contractor would want to know how many weeks in advance the contractor would need to place an order with the manufacturer.

In the construction stage, the contractor uses the schedules developed in the pre-construction planning to control the production process. These schedules need to be reviewed in line with the progress achieved, and updated at regular intervals.

The contractor manages date-to-date construction activities using short-term schedules. These are prepared each week and show in detail the work to be accomplished by committed resources.

The contractor also prepares a range of schedules for control of materials, labour, subcontractors, and plant/equipment. These will be discussed in more detail in Chapter 4.

1.6.5 Planning in the commissioning stage

The commissioning stage is reached at the end of the project construction stage, when the project is said to be 'practically complete'. The client may take possession of the project although it may not be fully completed. From the legal perspective, the project has reached the start of a defects liability period during which the contractor is required to fully complete any minor omissions of work under the contract, finalise commissioning of all the services, and repair any defects. It is also during this stage that the client occupies the project or tenants commence tenancy fitouts. Clearly, a schedule is needed at the commissioning stage to plan, organise and control such a wide range of activities. It is commonly the responsibility of the project manager to prepare such a schedule. Since it covers a period typically up to 12 months, the time-scale of the schedule is likely to be in weeks or days.

1.7 Examples of construction plans and schedules

There are many different types of plans and schedules used in construction planning and control. They vary from simple tables and charts to complex networks. The decision on what type of plan or schedule to use is generally governed by the size of a project, the WBS level of planning, which determines the level of detail and the time-scale, and the nature of the activities to be planned.

Plans and schedules must be presented in a manner that facilitates transfer of information from one party to another in the most effective manner. Too often they may be mathematically and technically correct but fail to communicate the planning information effectively. Furthermore, plans and schedules must be prepared in a manner suitable for monitoring and controlling of the production process.

The most common examples of plans and schedules used in the construction industry will now be briefly reviewed.

1.7.1 Lists

A list represents the simplest form of a schedule. It is nothing more than a written account of a series of activities presented in some logical order that need to be accomplished by a specific resource in a fairly short time, such as one day. The following example shows the simplicity of such a list.

A list of activities to be performed by the plastering crew on 28 June 2010:

- Sand back joints in plasterboard ceilings and walls in lunch and dining rooms
- Set joints (second coat) in plasterboard ceilings and walls in launch and dining rooms
- Fix plasterboard sheets to ceilings in bedrooms 1 and 2
- Fix plasterboard sheets to walls in bedrooms 1 and 2
- Tape and set joints in plasterboard ceilings and walls in bedrooms 1 and 2.

1.7.2 Tables

Tables are similar to lists but are presented in a more compact and systematic manner. For example, Table 1.1 shows specific activities that a site crane will perform on an hourly basis over one day.

A 'tender letting schedule' (TLS) is a matrix style table that is used to plan/track and monitor design and tender activities. The rows of the table are

Table 1.1 Specific activities of a site crane

Time	*Activity*	*Location*
7–8 am	Unloading trucks	Site loading zone
8–9 am	Lifting formwork	From level 2 to 6
9–10 am	Lifting formwork	From level 2 to 6
10–11 am	Pouring concrete	Level 5 columns
11–12 noon	Pouring concrete	Level 5 columns
12–1 pm	Lunch	
1–2 pm	Lifting reinforcement	From ground to level 6
2–3 pm	Lifting bricks	From ground to level 1
3–4 pm	Lifting electrical material	From ground to level 2
4–5 pm	Lifting a/c ducts	From ground to level 2

a/c: air conditioning

categorised into subcontract trade packages and the columns into the different steps of the tender letting process. These steps typically include:

- The site work commencement date(s)
- Procurement of materials, labour, plant, equipment
- The subcontract tender process steps
- Dates of design completion.

The TLS is typically prepared after the construction bar chart has been produced. This consequently results in the due dates for each step of the tender letting process being calculated working back from the site work commencement date(s), with the design completion date being the final calculated. Once the TLS has been developed, it is then reviewed by the design and contract administration (tender) teams for calculation of resources required to meet the schedule due dates. The 'due dates' for all activities are then set as a baseline tracking reference for the rest of the project. The TLS can be used as a management tool for monitoring and forecasting the design and tender activities, and forecasting potential risks to meeting the site work commencement date(s).

The TLS should be prepared and implemented at the beginning of the detailed design phase of the project lifecycle, with its use ending once all of the subcontract trade packages are let. When employed effectively, the TLS is a useful tool for controlling and reducing the risk of delays in the delivery phase of the project arising from the design and tender phases. An example of a TLS is shown at Figure 1.5.

1.7.3 Coloured or marked-up drawings

Contract drawings of large projects, particularly plans and sections, are often used to show daily or weekly expected progress by marking up the work to be

TENDER LETTING SCHEDULE FOR SUBCONTRACTS — Todays Date: 16-Feb-2011

Trade Package	Current Action By	ISSUE COMPLETE DESIGN	OUT TO TENDER	CLOSE TENDER	RECOM-MEND-ATION	HEAD OFFICE APPROVE	LET CONTRACT	PROCURE-MENT (In Weeks)	START ON SITE	NOTES
SITE ESTABLISHMENT										
Site Setup	Site								23-Feb-11	
Forecast Line		01-Jan-11	08-Jan-11	26-Jan-11	09-Feb-11	15-Feb-11	16-Feb-11	1	23-Feb-11	
Surveyor	H.O						16-Feb-11	1	23-Feb-11	
Forecast Line		03-Jan-11	10-Jan-11	31-Jan-11	14-Feb-11	20-Feb-11	20-Feb-11	2 days	23-Feb-11	Tender approval late by 5 days.
Traffic Management	Site								23-Feb-11	
Forecast Line		29-Dec-10	05-Jan-11	26-Jan-11	09-Feb-11	15-Feb-11	16-Feb-11	1	23-Feb-11	
GROUNDWORKS										
Civil Contractor	C.A				13-Feb-10	19-Feb-10	20-Feb-10	1	27-Feb-11	
Forecast Line		08-Jan-11	15-Jan-11	05-Feb-11	19-Feb-11	19-Feb-11	20-Feb-11	1	27-Feb-11	H.O approval reduced to 1 day.
Hydraulic Services	C.A	08-Mar-10	15-Mar-10	05-Apr-10	19-Apr-10	25-Apr-10	26-Apr-10	1	03-May-11	
Forecast Line		08-Mar-11	15-Mar-11	05-Apr-11	19-Apr-11	25-Apr-11	26-Apr-11	1	03-May-11	
Electrical Services	C.A	22-Mar-10	29-Mar-10	19-Apr-10	03-May-10	09-May-10	10-May-10	1	17-May-11	
Forecast Line		22-Mar-11	29-Mar-11	19-Apr-11	03-May-11	09-May-11	10-May-11	1	17-May-11	
SUBSTRUCTURE										
Formworker	DESIGN	23-Mar-11	30-Mar-11	20-Apr-11	04-May-11	10-May-11	11-May-11	2	25-May-11	
Forecast Line		23-Mar-11	30-Mar-11	20-Apr-11	04-May-11	10-May-11	11-May-11	2	25-May-11	
Reinforcement	DESIGN	28-Mar-11	04-Apr-11	25-Apr-11	09-May-11	15-May-11	16-May-11	2	30-May-11	
Forecast Line		28-Mar-11	04-Apr-11	25-Apr-11	09-May-11	15-May-11	16-May-11	2	30-May-11	
Concreter	DESIGN	04-Apr-11	11-Apr-11	02-May-11	16-May-11	22-May-11	23-May-11	2	06-Jun-11	
Forecast Line		04-Apr-11	11-Apr-11	02-May-11	16-May-11	22-May-11	23-May-11	2	06-Jun-11	
Structural Steel	C.A	31-Mar-11	07-Apr-11	28-Apr-11	12-May-11	18-May-11	20-May-11	6	01-Jul-11	
Fcst Line		01-Apr-11	08-Apr-11	29-Apr-11	13-May-11	19-May-11	20-May-11	6	01-Jul-11	
Fire Proofing	DESIGN	09-Jun-11	16-Jun-11	07-Jul-11	21-Jul-11	27-Jul-11	28-Jul-11	1	04-Aug-11	
Forecast Line		09-Jun-11	16-Jun-11	07-Jul-11	21-Jul-11	27-Jul-11	28-Jul-11	1	04-Aug-11	

PAGE 1 of 6

Baseline Planned 'Due Dates'

Forecast 'Due Dates'

Completed Activities Shaded

Late Dates Highlight in Colour

Figure 1.5 Example of a tender letting schedule.

performed using different colours or textures. Such plans are highly visual and are useful in planning activities such as bulk excavation, installation of services, horizontal formwork and the like.

1.7.4 Diagrams

Flow charts, bar charts, line of balance charts and multiple activity charts are common examples of diagrams.

A 'flow chart' is a simple diagram showing a flow of information or work. It is particularly useful for developing a logical sequence of activities and for illustrating the flow of work in the production process. A flow chart is shown in Figure 11.6.

A 'bar chart' is the most commonly used scheduling tool. It is relatively easy to prepare and read, but its clarity diminishes with an increasing number of activities. An example of a bar chart is given in Figure 2.2.

A 'multiple activity chart' is used effectively for allocating the work to committed resources on daily basis; see Figure 9.5.

A 'line of balance' is a chart used for scheduling highly repetitive work, which is performed by different resources; see Figure 10.7.

A 'staging diagram' is used for graphically representing high-level project strategies. The information displayed on the staging diagram is derived from the construction bar chart schedule. Staging diagrams usually contain a series of pages, each representing the planned project strategy at each phase. A staging diagram is shown in Figure 1.6.

A 'procurement schedule' is a variation of the bar chart which tracks the supply chain process of materials and equipment that have a long lead-time (typically over four weeks). The procurement schedule is quick and easy to produce but difficult to amend. A procurement schedule is shown in Figure 1.7.

1.7.5 Graphs

Graphs are useful to highlight the relationship between two or more variables. They are used in many applications including cash flow forecasting, assessing distribution of resources, production planning and analysing production trends; see Figure 6.3.

1.7.6 Networks

Networks are diagrammatic models showing a logical sequence and relationships among activities. They provide a basis for the critical path and the PERT (Program Evaluation and Review Technique) methods of scheduling. These will be discussed in detail in Chapters 3 and 13 respectively. An example of the network is given in Figure 3.4.

The advantages and disadvantages of different presentation forms of plans and schedules are summarised in Table 1.2.

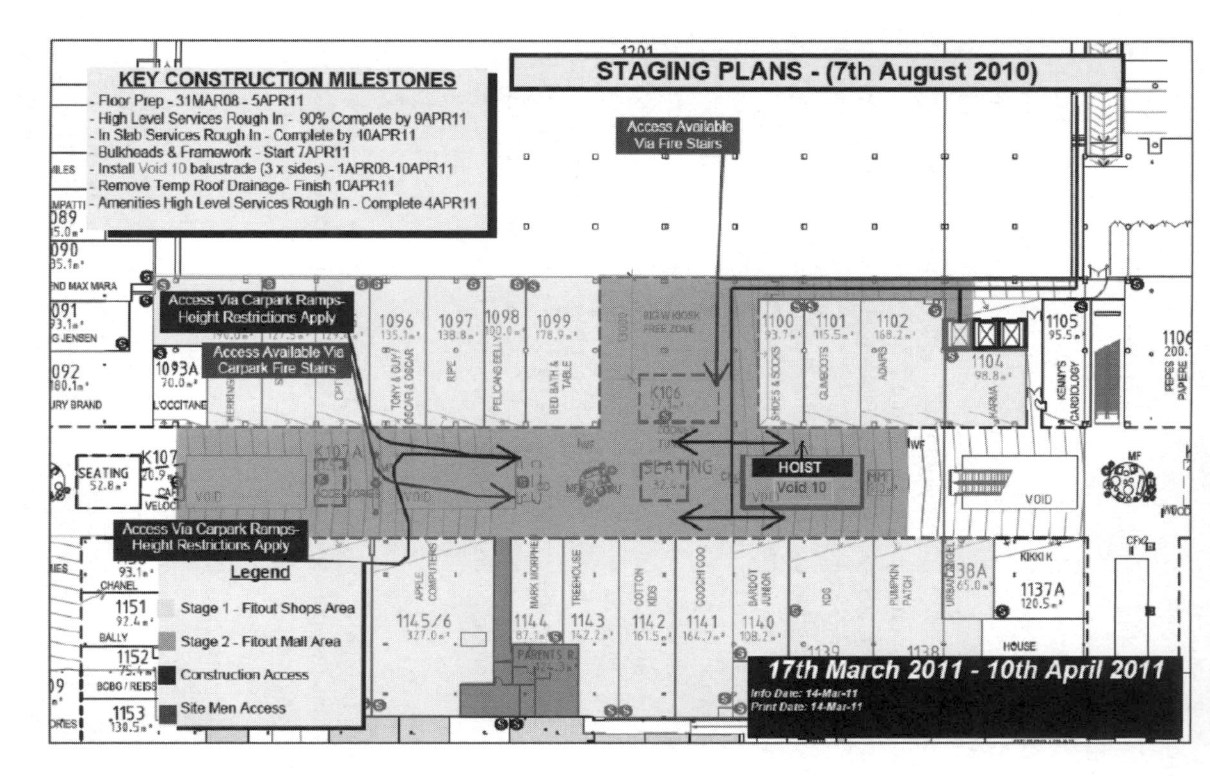

Figure 1.6 Example of a staging diagram.

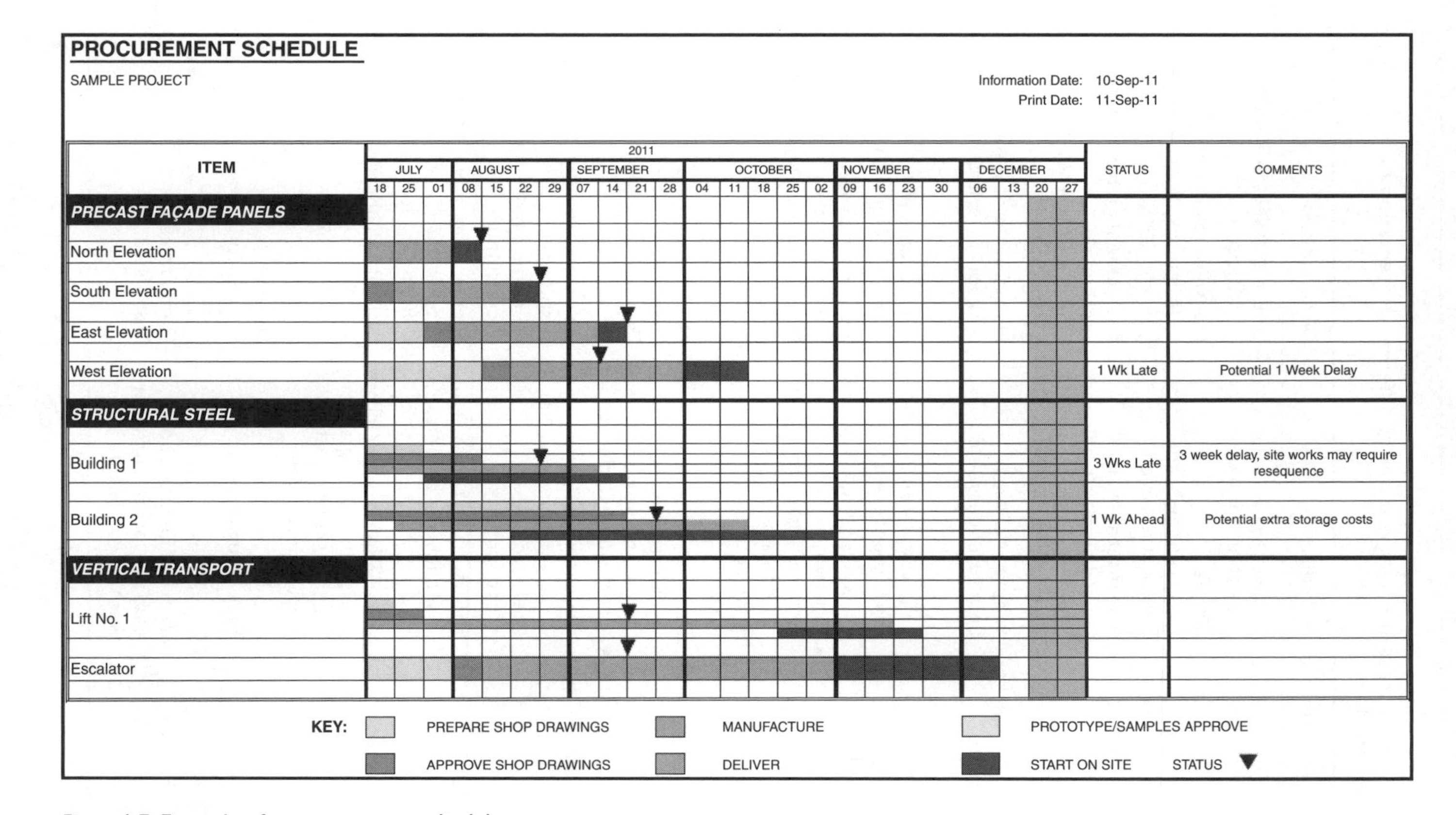

Figure 1.7 Example of a procurement schedule.

Table 1.2 Advantages and disadvantages of different presentation forms of plans and schedules

Presentation forms	*Advantages*	*Disadvantages*
List	Simple Easy to understand	Doesn't reflect complex situations
Table	Simple Easy to understand Better structured	Doesn't reflect complex situations
Tender letting schedule	A management tool used to plan/track and monitor design and tender activities	Supplements a bar chart schedule Doesn't reflect complex situations
Coloured or marked-up drawing	Simple and visual Easy to understand	Not capable of time and resource scheduling
Diagrams		
Flow chart	Simple and visual Easy to understand	Not capable of time and resource quantification
Bar chart	Shows sequencing of activities Generally easy to understand Time-scaled Visual	Relationships between activities shown by links Not capable of quantification
Line of balance	Shows a delivery programme Simple and powerful Visual when coloured	Not supported by computer software Difficult to update
Multiple activity chart	Allocates work to committed resources Useful to assess productivity	Not supported by computer software Its development is time-consuming
Staging diagram	Shows project phases and sequencing graphically	Supplements a bar chart schedule Difficult to update
Procurement schedule	A variation of a bar chart used in tracking supply chain activities of 'long lead-time items'	Difficult to update
Graph	Simple Easy to understand	Doesn't reflect complex situations
Networks		
CPM and PERT	Work sequence shown Easy to update and alter Resources considered	Complicated Not visual

1.8 Summary

This chapter described the planning process in broad terms and identified important issues relevant to planning and controlling construction projects. In particular, it examined the hierarchy of schedules defined by the WBS, types of schedules, planning tasks in different stages of the project lifecycle, and examples of plans and schedules commonly used in the construction industry. In the next chapter, the scheduling technique known as a bar chart will be discussed in detail.

Chapter 2

Bar charts

2.1 Introduction

The purpose of this chapter is to examine bar chart scheduling technique. A bar chart structure will be defined first, followed by the description of the logical steps taken in developing a bar chart. The development and application of method statements, which serve the purpose of recording planning information for future use, will be examined next, followed by a brief discussion on limitations of traditional, unlinked bar charts. The importance of activity links in bar charts will be emphasised. Finally, the use of bar charts will be demonstrated on a simple example.

The 'bar chart' is the first scheduling technique that will be examined in this book. It is most popular and is widely used in the construction industry. The work of Henry L. Gantt and Frederick W. Taylor in the early 1900s, which was associated with graphic representation of work on a time-scale, led to the development of the Gantt, or bar, chart. The term 'bar chart' is more commonly used and will be adopted in this book.

Scheduling is a decision-making task. It is also a modelling task performed on the model provided by the format of the bar chart. In scheduling, the planner reduces the physical size of the project to a relatively small number of activities that can fit a sheet of paper. Upon finding the best solution, the planner relates it to the actual project. In any modelling exercise, the accuracy of the solution depends on the development of a model that is representative of the actual project.

2.2 What is a bar chart?

A bar chart is a simple, visual scheduling tool that is easy to use. It displays planning information graphically in a compact format to a time-scale. It is a diagram divided into columns and rows. Columns represent a given time-scale, which could be expressed in months, weeks, days or even hours. Activities are scheduled as bars within horizontal rows.

The first column lists activities that are to be scheduled in more or less the logical order of production. The production process is then represented by

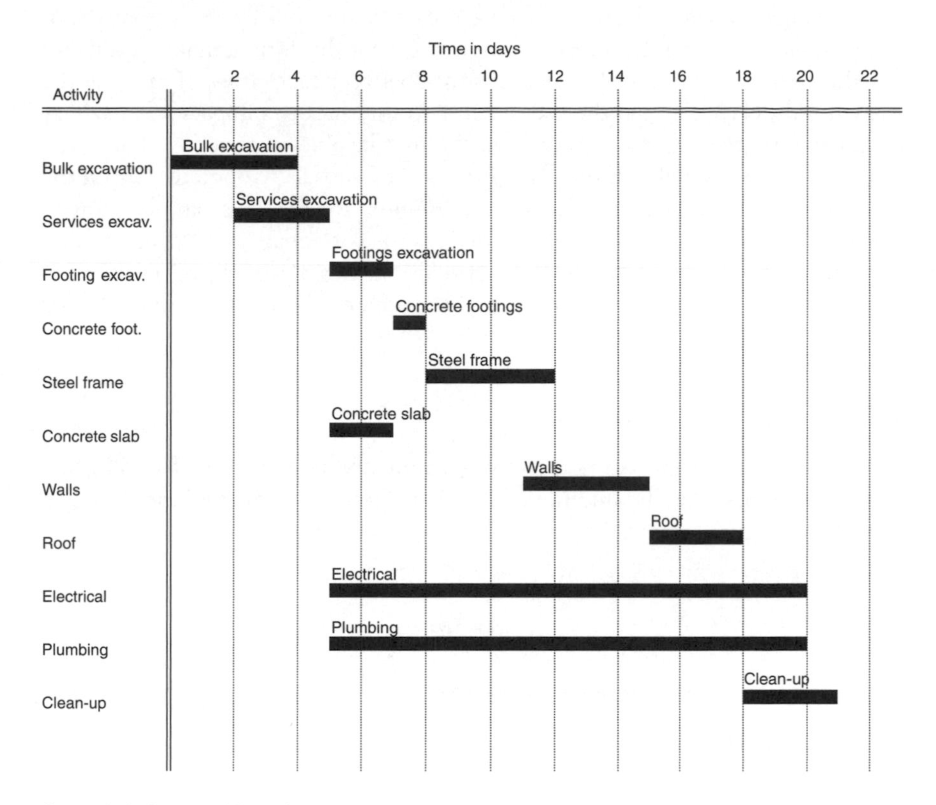

Figure 2.1 A typical bar chart.

horizontal bars which are drawn for each activity within the time-frame of the bar chart. The length of an activity bar gives activity duration. Figure 2.1 is an example of a simple bar chart.

The start and end points of an activity bar are significant in determining the position of that activity within a logical production sequence. In other words, the start point of an activity bar is closely related to the end point of a preceding activity bar. Similarly, the end point of an activity bar shows the relationship between that activity and the following activities. For example, the start of the activity 'Footings excavation' is linked to the finish of the preceding activity, 'Services excavation', and the finish of the activity 'Footings excavation' is linked to the start of the next activity, 'Concrete footings'.

However, it is unclear whether the start of the activity 'Concrete footings' is related only to the finish of the preceding activity, 'Footings excavation', or to 'Concrete slab' as well. It is also difficult to determine whether the activity 'Concrete footings' is in any way related to the activities 'Electrical' and 'Plumbing'.

The graph shown in Figure 2.1 represents a traditional format of bar chart. It should be clear from the foregoing discussion that this format makes it difficult for the user to interpret the relationships between activities. If one of the activities is delayed, would the user be able to interpret the impact of this delay on other activities? For example, would the user be able to correctly deduce the impact of a one-day delay in the completion of the activity 'Concrete slab' on the activities 'Steel frame', 'Electrical' and 'Plumbing'? For this simple project, the experienced user would most likely deduce the correct answer. When projects become more complex, interpretation of relationships among activities is much more difficult. This problem can easily be overcome by constructing a linked bar chart.

2.3 Linked bar chart

When the end of a preceding activity is connected to the start of a following activity by a link line, the traditional bar chart format is converted into a linked format (see Figure 2.2).

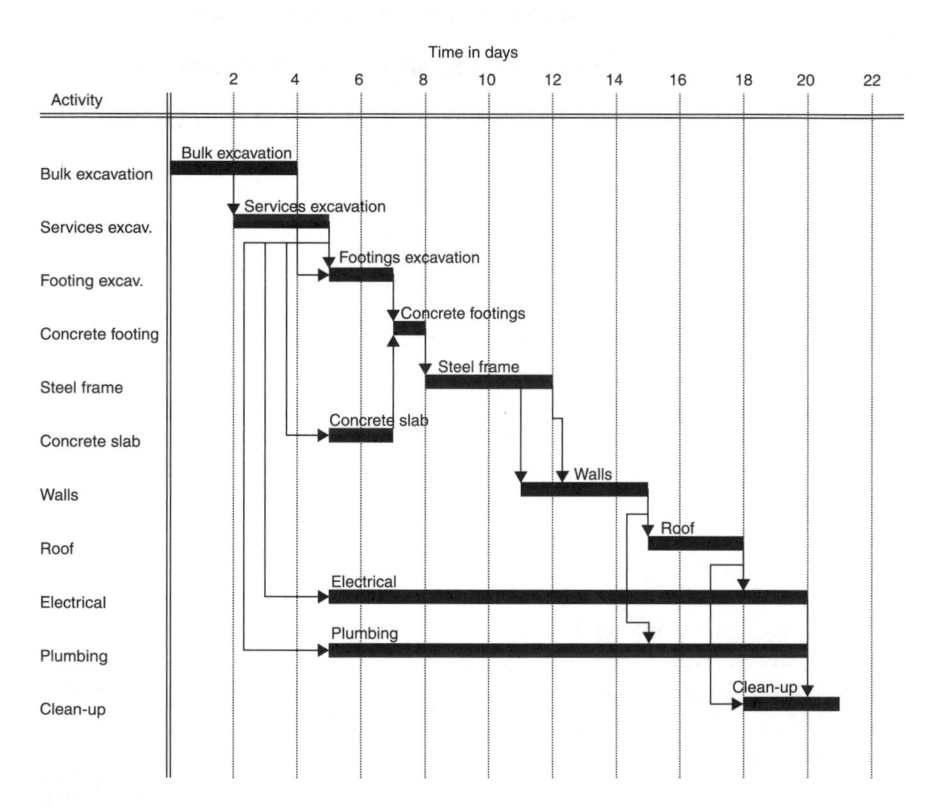

Figure 2.2 A linked bar chart.

A linked bar chart provides a clear picture of relationships among scheduled activities. It even defines relationships between those activities that are scheduled concurrently. For example, the completion of 'Walls' affects the completion of 'Plumbing'; similarly, the completion of 'Roof' affects the completion of 'Electrical'. Without these links, the relationship between the activities 'Wall' and 'Plumbing', and 'Roof' and 'Electrical' would be extremely difficult to define.

Linking of activities has overcome the main shortcoming of bar charts. Positive features of a bar chart scheduling technique include:

- Speed and ease of development
- Ease of understanding
- The ability to schedule complex relationships among activities
- The ability to communicate information
- The ability to monitor and control the production process.

Despite numerous benefits, linked bar charts are rarely used as a primary planning tool, particularly on larger projects. This is because they lack a computational base, which prevents calculation of a critical path (for the definition of a critical path, see Chapter 3). It also prevents periodic recalculation of the schedule, which is necessary for regular updating.

2.4 Process of developing a bar chart

A typical planning process is defined in Chapter 1. It provides a framework on which a process of developing a bar chart or any other technique used in scheduling is based. Logical steps in developing a bar chart include:

1. Identifying the work to be done and setting an objective
2. Determining the extent of planning detail for a particular level of work breakdown structure (WBS)
3. Breaking the work down into activities
4. Developing alternative planning strategies:

   i. Preparing a logic diagram (that is, a logical sequence of activities) for each alternative
   ii. Determining duration of activities (see Chapter 11 for more details) based on the volume of work and the required resources (that is, people, plant/equipment, materials)
   iii. Recording planning information in method statements (see section 2.7 for details)
   iv. Preparing preliminary bar charts, one for each alternative
   v. Considering the use of resources; for example, one crew of workers cannot be scheduled to work on two separate activities at the same time

   vi Checking the volume of resources at each time interval to prevent their unnecessary accumulation.

5 Selecting the preferred planning strategy
6 Reviewing the preferred planning strategy illustrated on a bar chart schedule. Does it all make sense? Will the user be able to understand it? Does the bar chart schedule include enough information to be workable? Has anything been left out? Is there a better alternative? Does the bar chart schedule meet the planning objectives?
7 Committing to the bar chart schedule
8 Monitoring the progress regularly.

2.5 Activity duration

An activity is a task that needs to be accomplished. It describes a particular type of work, for example bulk excavation or plumbing. But 'work' may not always involve human activity; for example the curing of concrete occurs by natural means. Nevertheless, it must be included in a schedule since it adds time to the project. The issue of activity duration will be discussed in Chapter 11.

2.6 Risk contingency

Schedules built up from 'average' estimates of activity durations do not reflect the presence of risk. Risk that may cause delays in execution of the work is commonly assessed separately by the planner. Most frequent risks responsible for delays include inclement weather, latent site conditions, variations orders, unavailability of resources, re-work, accidents and the like. From the contractor's perspective, delays caused by the client are not risk events if the contractor is able to claim time extension under the contract, for example delays caused by variations orders or latent site conditions. For such risk events, the contractor will add no time contingency to the schedule. However, the other risks for which the contractor is responsible would need to be carefully assessed and added to the schedule in the form of a time contingency. The contractor may deal with a time contingency in one of the following ways:

- Add time contingencies to 'risky activities' only, or
- Add a time contingency to the whole project as a lump-sum allowance (to the end date of the schedule), or
- Break up the lump-sum time contingency into a number of smaller contingencies that are then added to the schedule at regular intervals, for example each month.

The issue of contingency will be discussed in Chapters 8 and 12.

2.7 Method statement

In determining the duration of activities, the planner makes a series of decisions about required resources, productivity and output rates of resources, and assessment of risk. If unrecorded, these important decisions would be largely unknown to those who are responsible for the schedule's implementation. Instead of using the previously formulated decisions, they would need to replicate it entirely.

A detailed account of decisions made by the planner in the preparation of a schedule can effectively be compiled in the form of a 'method statement'. When compiled systematically for each activity, method statements store highly detailed information of all key scheduling decisions for future retrieval. The project manager is then able to review information in the method statements and use it in developing the final operational schedule that meets all the contract requirements.

The format and the development of method statements will now be briefly demonstrated on the following simple example.

Let's assume that a contractor's planner is developing a tender schedule for construction of a single story storage facility. The work involves:

- Excavating the site
- Excavating footings
- Forming footings
- Reinforcing footings
- Concreting footings
- Stripping footings' formwork
- Forming the ground floor slab (edge formwork only)
- Reinforcing the ground floor slab
- Concreting the ground floor slab
- Forming walls (internal formwork)
- Forming walls (external formwork)
- Reinforcing walls
- Concreting walls
- Stripping walls
- Erecting the steel roof
- Installing roofing
- Electrical work
- Cleaning up.

Durations of activities are calculated from the quantities of work, and productivity and output rates of required resources. A method statement in Table 2.1 gives information on quantities, units of quantities, output rates of specific resources, number of resources, duration and the actual output rates. A largely sequential bar chart schedule for this project is given in Figure 2.3.

Table 2.1 A method statement

	Quantity	*Unit*	*Resource output per day*	*Number of resources*	*Duration (days)*	*Actual total output*
Excavate site	2,000	m^3			5	
1 m^3 track excavator			450	1		2,225
Labour to assist				1		
Excavate footings	100	m^3			2	
1/3 m^3 backhoe			80	1		160
Labour to assist				1		
Form footings	320	m^2			5	
Formworkers			70	4		350
Reinforce footings	6	Tonne			4	
Steelfixers			1.5	6		6
Concrete footings	60	m^3			1	
Concrete pump			180	1		180
Concreters			100	6		100
Strip footing formwork	320	m^2			2	
Formworkers			160	6		320
Form ground floor (edges)	160	m			2	
Formworkers			100	8		200
Reinforce ground floor slab	22	Tonne			7	
Steelfixers			3.5	6		24.5
Concrete ground floor slab	240	m^3			2	
Concrete pump			180	1		360
Concreters			140	8		280
Form walls (internal)	480	m^2			6	
Formworkers			90	10		540
Form walls (external)	490	m^2			6	
Formworkers			90	10		540
Reinforce walls	10	Tonne			7	
Steelfixers			1.5	6		10.5
Concrete walls	110	m^3			2	
Concrete pump			180	1		360
Concreters			60	6		120
Strip walls	980	m^2			5	
Formworkers			200	6		1,000
Erect steel roof	50	Tonne			7	
Mobile crane			40	1		280
Riggers			8	6		56
Install roofing	160	m^2			6	
Mobile crane			100	1		600
Riggers			30	6		180
Electrical	Item				43	
Subcontracted			–	2		–
Clean-up	Item				4	
Labourers			–	6		–

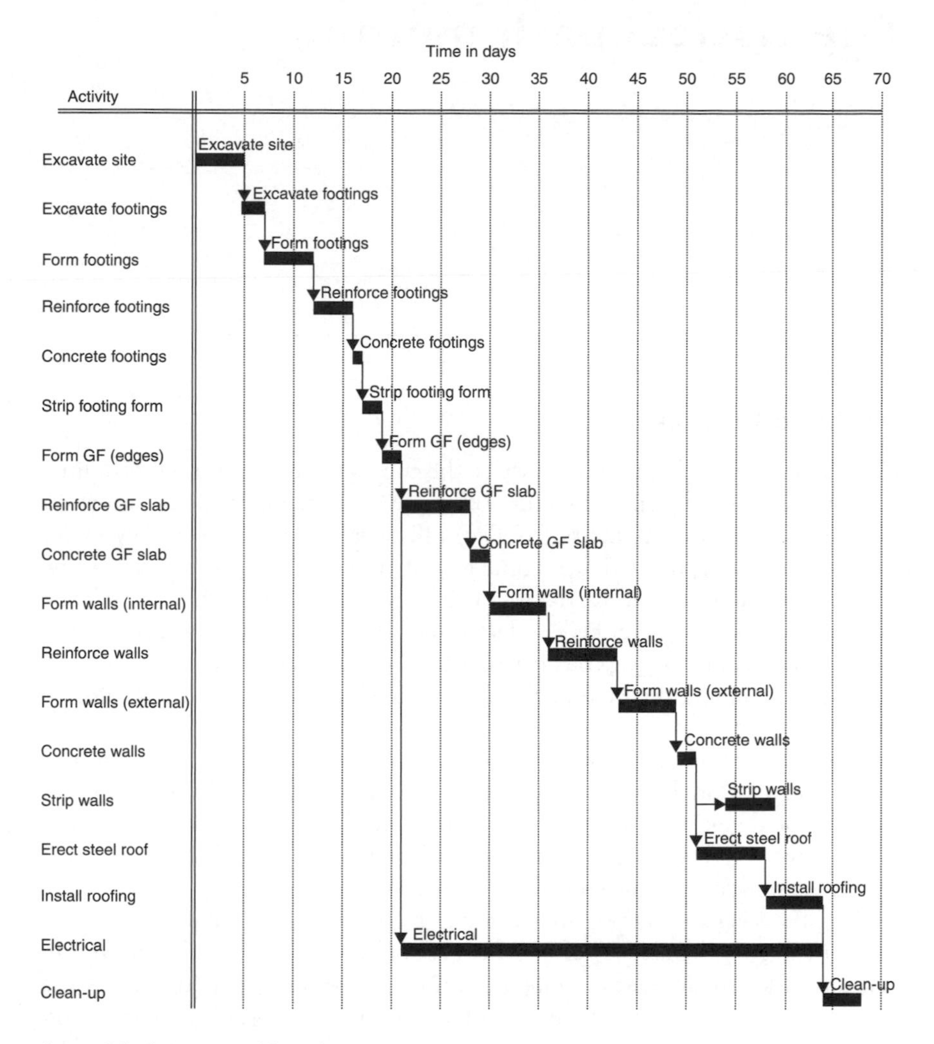

Figure 2.3 A sequential bar chart.

2.8 Summary

This chapter has reviewed the fundamentals of bar chart scheduling. In particular, it focused on defining a process of developing a bar chart and explaining the difference between the traditional and the linked formats of bar charts. The chapter also introduced the process of determining activity duration. In the final section, the importance of method statements in recording planning information was discussed. The next chapter will examine the fundamentals of critical path scheduling.

Chapter 3

The critical path method

3.1 Introduction

The purpose of this chapter is to describe the concept of critical path scheduling, which is the most frequently used time-scheduling technique in the construction industry. The critical path method (CPM) will be defined first, followed by a brief history of its development. The original format of the CPM, known as the 'arrow method', will only briefly be discussed. It is a later version, the 'precedence method', that will be described in detail, including calculation of forward and backward paths, identifying the critical path, and the computation of 'float'. An alternative computational method known as the 'link lag method' will also be presented.

Although a bar chart is effective in communicating planning information, it lacks a computational base for more detailed analysis of a schedule. The lack of a computational algorism prevents the planner from locating a critical path (that is, the longest path through a schedule), calculating float, analysing the use of resources, and accelerating a schedule at the least possible cost.

The CPM overcomes these shortcomings by providing a computational algorism for a highly detailed level of schedule analysis. Complex calculations are now universally performed by computers; however, an experienced scheduler is needed for the data input including the formulation of a logic of the production process and for the analysis of the output. Given that the output is constrained by the parameters within which the computer software operates, the planner is not only required to understand such parameters but is also expected to have sound knowledge of the theory of the CPM to be able to interpret a computer generated output and formulate an appropriate solution. For this reason, the theory of the CPM will be presented first in Chapters 3–6. Scheduling by computer will then be discussed in Chapter 7.

3.2 The critical path method

The aim of the CPM is to find a specific chain of activities that cannot be delayed without delaying the end date of the schedule. The path connecting these activities from the start to the end of the schedule is known as a 'critical path' and the activities are 'critical activities' in the sense that if any one of them is delayed, the entire schedule would be delayed. Non-critical activities, on the other hand, may be delayed, if needed, by the amount of time known as 'float' (see section 3.5.2 for definition). The CPM also provides a framework for a highly detailed analysis of committed resources and for achieving the most effective schedule compression by keeping the cost and time at optimum. These issues will be discussed in detail in Chapters 4 and 6 respectively.

A CPM schedule is developed in the form of a graphic model known as a 'network'. A network is a maze of activities linked together in a logical sequence to create a visual map of relationships and dependencies. A convention has been established for representing activities in networks. In the arrow network, activities are shown as arrows, each accompanied by a pair of circles, while in the precedence network, activities are expressed as a set of connected boxes.

3.3 A brief history of the critical path method

The CPM was first used in Great Britain in the mid-1950s on the construction of a central electricity-generating complex. Its full potential was later realised by Walker of Du Pont and Kelley of Remington Rand, in the USA. Their CPM was based on a graphic network commonly referred to as the 'arrow method'. It was driven by a computational process requiring no more than additions and subtractions. The benefits of critical path scheduling were quickly realised by a wide range of organisations including construction firms, many of whom have successfully implemented it in their planning. The advent of the CPM computer software in the 1970s has made the CPM a universal scheduling technique.

In 1961, Professor Fondahl of Stanford University presented a different version of the CPM, known as the 'precedence method'. Fondahl's method offers a number of improvements over the arrow method, particularly with regard to schedule construction and its analysis. This method has become the preferred CPM.

The CPM is widely used throughout the construction industry. Apart from contractors, who are the main proponents of the method, other project participants such as clients, designers, consultants and subcontractors rely on the CPM in scheduling. In addition to planning activities, the CPM method assists in organising resources and controlling progress.

3.4 The arrow method

Although the arrow method of critical path scheduling has largely been superseded by the precedence method, it is nevertheless useful in just a few words to distinguish it from the precedence method.

The arrow method of critical path scheduling is 'event-oriented' because it emphasises individual events rather than activities. Events are shown as nodes or circles and signify start and finish points of activities. An activity is shown as an

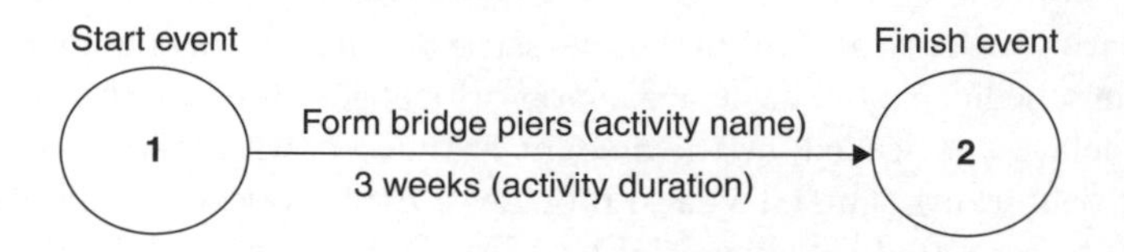

Figure 3.1 Graphic representation of activity in an arrow schedule.

arrow that connects a pair of start and finish events of that activity. The length of an arrow has generally no significance since arrow networks are not drawn to a time-scale. Events have no time duration but activities do. Figure 3.1 shows how an activity is represented by its two events and an arrow.

The activity 'Form bridge piers' in Figure 3.1 is interpreted in the following terms:

- Event 1 is the start of the activity 'Form bridge piers'.
- Event 2 is the finish of the activity 'Form bridge piers' and the start of another event.
- An arrow connecting events 1 and 2 is the activity 'Form bridge piers'.

Finish events of preceding activities are linked to start events of succeeding activities to form relationships depicting the logic of a production sequence. The convention requires arrows to be drawn from left to right in the arc between horizontal and vertical to ensure that for any pair of preceding and succeeding activities, the preceding activity is to the left of the succeeding activity. Drawing arrows from right to left is not allowed.

When activity A in Figure 3.2 is followed by activity B, only one event is needed to show the end of activity A and the start of activity B. The finish event of activity A and the start event of activity B merge into one (the circled 2). Similarly, other finish events of preceding activities and start events of succeeding activities will merge.

When an arrow schedule is completed, its events are numbered from left to right, ensuring that the number of the start event is smaller than the number of the finish event. A pair of start and finish event numbers then represents the activities to be performed in a schedule. For example, activity B in Figure 3.2 is identified as activity 2–3. Activities need to be identified by event numbers to facilitate computer processing.

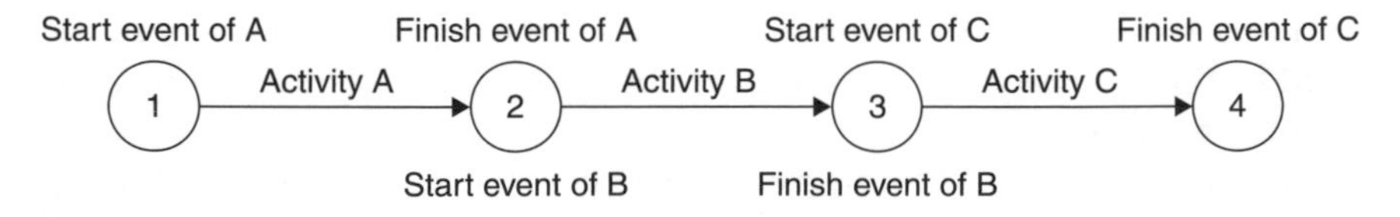

Figure 3.2 Example of a chain of activities in an arrow schedule.

An inherent problem with the arrow method is that construction of more complex schedules requires careful separation of preceding and succeeding activities at various events to avoid creating unwanted dependencies amount activities. This requires the planner to insert 'dummy activities', i.e. activities that have no duration, into a schedule. In comparison, the later method known as the precedence method is simple in its structure and requires no 'dummy activities' to depict the correct logic of a schedule. This method will now be discussed in detail.

3.5 The precedence method

The precedence method is activity-oriented. There are no events in the precedence method. A square or a box graphically represents activities, though other shapes such as a circle or a hexagon have also been used. The Standards Australia HB 24–1992: *Handbook symbols and abbreviations for building and construction* (AS 1992) prescribes a box for representing activities in a precedence schedule.

The protocol shown in Figure 3.3 has been adopted in this book for labelling activity boxes in a precedence schedule.

The construction of a precedence network is fairly simple. Since the precedence method is activity-oriented, relationships among activities are formed by simply linking activities together, and these links are referred to as dependency lines. The flow of work is shown in a schedule to progress from left to right. Dependency lines may be drawn up and down, but never from right to left. For ease of interpreting a schedule, it is suggested that arrowheads be attached to dependency lines to show the direction of the workflow.

For the time being the schedule construction will be based on the assumption that the preceding activity must be fully completed before the succeeding activity could begin. This assumption will be removed in Chapter 5 when different overlapping models will be introduced.

The manner in which activities in a precedence schedule are linked together is illustrated in Figure 3.4. The schedule shows the following relationships:

- Activities A, B and C are the independent start activities
- Activity D cannot start until activities A and C have been completed

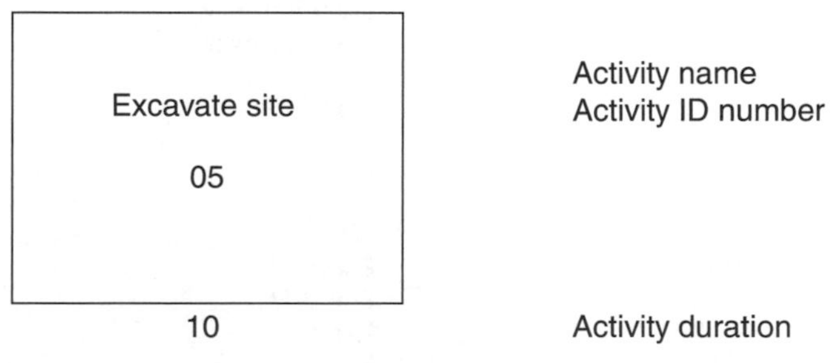

Figure 3.3 Labelling protocol.

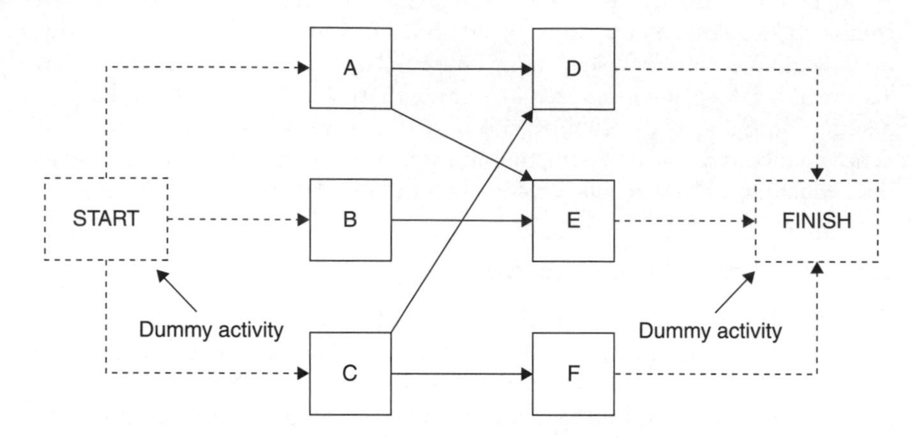

Figure 3.4 Linking of activities in a precedence schedule.

- Activity E cannot start until activities A and B have been completed
- Activity F cannot start until activity C has been completed
- Activities D, E and F are the independent end activities.

Dummy activities 'Start' and 'Finish', identified by the dashed outlines, may be used to show the start and finish points of a schedule. This is illustrated in Figure 3.4. The dummy activities 'Start' and 'Finish' have zero duration and consequently add no time to the schedule. Dependency lines originating from or leading to 'Dummy Activities' are shown as dashed lines.

Let's construct a precedence schedule based on information in Table 3.1. No matter how complex relationships among the activities are, the defined

Table 3.1 Typical information required for construction of a precedence network

Activity	*Time duration (days)*	*Depends on completion of:*
A	4	Start activity
B	2	Start activity
C	3	Start activity
D	2	A
E	1	B and D
F	4	C
G	4	D
H	3	F
I	5	G
J	3	E and H
K	6	E and H
L	9	I and J
M	5	K

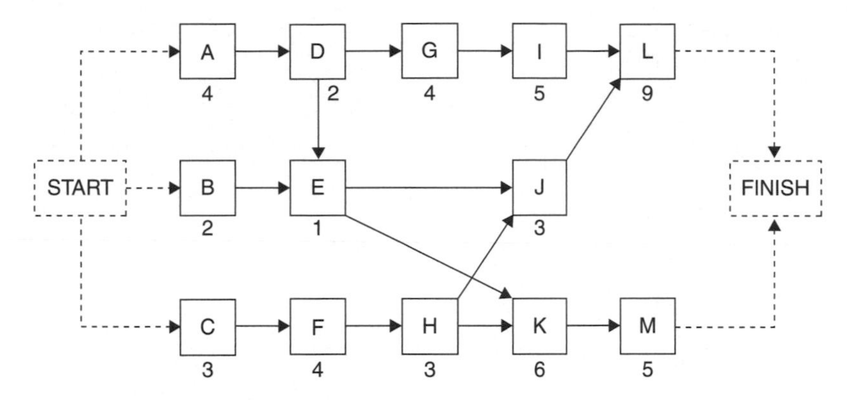

Figure 3.5 Example of a precedence schedule.

relationships are easily constructed by simply linking the activities together. The final shape of a schedule may vary depending on the planner's personal preference. One possible configuration of a schedule is shown in Figure 3.5.

The dummy activity 'Start' defines the start point of the schedule, which has three start activities, A, B and C. Similarly, the dummy activity 'Finish' defines the end point of the schedule for two end activities, L and M.

3.5.1 The critical path

The critical path is the longest path through a schedule (from the start activity to the finish activity). A typical schedule will have many paths, of which at least one will be critical. A schedule may have more than one critical path and in the most extreme case all of its paths may in fact be critical.

The knowledge of the location of critical activities is vital for the effective planning and control of a project. It assists the planner in allocating sufficient resources to critical activities to ensure their completion on time.

The calculation process of the critical path schedule involves two distinct steps: a forward pass and a backward pass through a network. A 'forward pass' calculates earliest start date (ESD) and earliest finish date (EFD) of activities by working through a schedule from its start to its finish. A 'backward pass' then calculates latest start date (LSD) and latest finish date (LFD) of activities by working through a schedule from its finish to its start.

The convention for recording scheduling information in a precedence network is given below. It will be used throughout this book.

For clarity, 'FF' and 'TF' are the abbreviations for 'free float' and 'total float'. Their definitions and significance will be discussed in section 3.5.2.

When a critical path schedule is calculated manually, the ESD of the first activity in a schedule is assumed for simplicity's sake to be zero, but CPM computer software defines the ESD of the project as week 1, day 1, hour 1.

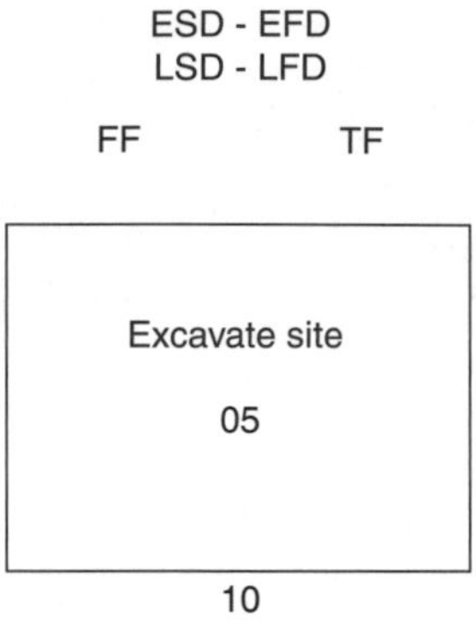

Figure 3.6 Labelling protocol.

For a schedule with only two activities, I and J, the ESD of the preceding activity I is assumed to be zero. Its EFD is calculated using the following general formula:

$$EFD_I = ESD_I + Duration_I$$

Let's proceed with a *forward pass* using a schedule in Figure 3.5. The ESD of the first dummy activity 'Start' is assumed to be zero. Because a dummy activity has zero duration, the EFD of the dummy activity 'Start' must also be zero.

Since the start activities A, B and C cannot start until the dummy activity 'Start' has been completed, the ESD of activities A, B and C will start when the preceding dummy activity 'Start' has been completed. It follows that activities A, B and C will all have the ESD as zero. The EFD of each of those activities is then calculated as follows:

$$EFD_A = 0 + 4 = 4$$

$$EFD_B = 0 + 2 = 2$$

$$EFD_C = 0 + 3 = 3$$

Given the assumption that the preceding activity must be fully completed before the succeeding activity could begin, the EFD of activity A becomes the ESD of activity D. Similarly, the EFD of activity C becomes the ESD of activity F. Therefore,

$$EFD_D = 4 + 2 = 6$$

$$EFD_F = 3 + 4 = 7$$

Activity E cannot start until its two preceding activities, B and D, have been completed. With regard to the link D–E, the ESD of activity E would be 6 (because $EFD_D = 6$) and with regard to link B–E, the ESD of activity E would be 2 (because $EFD_B = 2$). The earliest possible start date of activity E is 6 as defined by the link D–E. It cannot be 2 because neither of the preceding activities A and D would be fully completed. In general, when two or more preceding activities, $I_1, I_2 \ldots I_n$, join the same succeeding activity, J, the ESD of activity J will be the greatest or maximum value of EFDs of the preceding activities $I_1, I_2 \ldots I_n$. Consequently, the general formula is:

$$ESD_J = MAX\ EFD_{I_1, I_2 \ldots I_n}$$

Then,

$$EFD_E = 6 + 1 = 7$$

Calculations of ESDs and EFDs are shown in Figure 3.7. The remaining forward pass calculations are as follows:

$$ESD_G = EFD_D = 6, \text{ therefore } EFD_G = 6 + 4 = 10$$

$$ESD_H = EFD_F = 7, \text{ therefore } EFD_H = 7 + 3 = 10$$

$$ESD_I = EFD_G = 10, \text{ therefore } EFD_I = 10 + 5 = 15$$

$$ESD_J = EFD_H = 10, \text{ therefore } EFD_J = 10 + 3 = 13$$

$$ESD_K = EFD_H = 10, \text{ therefore } EFD_K = 10 + 6 = 16$$

$$ESD_L = EFD_I = 15, \text{ therefore } EFD_L = 15 + 9 = 24$$

$$ESD_M = EFD_K = 16, \text{ therefore } EFD_M = 16 + 5 = 21$$

The ESD of the dummy activity 'Finish' is 24 (governed by the EFD of activity L) and its EFD is also 24 since its duration is zero. This is the end of the forward pass computation.

The *backward pass* computation begins from the end activity in a schedule and proceeds to the start activity. It requires LSDs and LFDs to be calculated. For any activity I, the backward pass computation determines its LFD value first, from which the LSD value is calculated as follows:

$$LSD_I = LFD_I - Duration_I$$

In Figure 3.7 the EFD of the dummy activity 'Finish' determines the overall schedule duration, which is 24. Since it is the last activity, its LFD must also be 24. Therefore, its LSD is 24 minus zero, which is 24. The LSD of the dummy activity 'Finish' becomes the LFD of the preceding activities L and M. Then,

$$LFD_L = 24$$

$$LFD_M = 24$$

Consequently,

$LSD_L = 24 - 9 = 15$

$LSD_M = 24 - 5 = 19$

Similarly,

$LFD_I = LSD_L = 15$, therefore $LSD_I = 15 - 5 = 10$

$LFD_J = LSD_L = 15$, therefore $LSD_J = 15 - 3 = 12$

$LFD_K = LSD_M = 19$, therefore $LSD_K = 19 - 6 = 13$

$LFD_G = LSD_I = 10$, therefore $LSD_G = 10 - 4 = 6$

When two or more succeeding activities, $J_1, J_2 \ldots J_n$, originate from the same preceding activity, I, the LFD of activity I will be the minimum LSD time of the following $J_1, J_2 \ldots J_n$ activities. The general formula is:

$$LFD_I = \text{MIN } LSD_{J_1, J_2 \ldots Jn}$$

For example, the succeeding activities, J and K in Figure 3.7, have a preceding link with activity E. Since the LSD of activity J is smaller than that of activity K, the LFD of activity H will be equal to the ESD of activity J. The remaining calculations are then as follows:

$LFD_H = LSD_J = 12$, therefore $LSD_H = 12 - 3 = 9$

$LFD_E = LSD_J = 12$, therefore $LSD_E = 12 - 1 = 11$ (activity E has two succeeding activities, J and K)

$LFD_D = LSD_G = 6$, therefore $LSD_D = 6 - 2 = 4$ (activity D has two succeeding activities, E and G)

$LFD_F = LSD_H = 9$, therefore $LSD_F = 9 - 4 = 5$

$LFD_A = LSD_D = 4$, therefore $LSD_A = 4 - 4 = 0$

$LFD_B = LSD_E = 11$, therefore $LSD_B = 11 - 2 = 9$

$LFD_C = LSD_F = 5$, therefore $LSD_C = 5 - 3 = 2$

$LFD_{START} = LSD_A = 0$, therefore $LSD_{START} = 0 - 0 = 0$ (dummy activity 'Start' has three succeeding activities, A, B and C)

The precedence network in Figure 3.7 includes both the forward and backward paths calculations. It also highlights critical activities, which are identified by having the same ESD and LSD times (as well as EFD and LFD times). The critical activities A, D, G, I and L then form a critical path, which is the longest pass through the network.

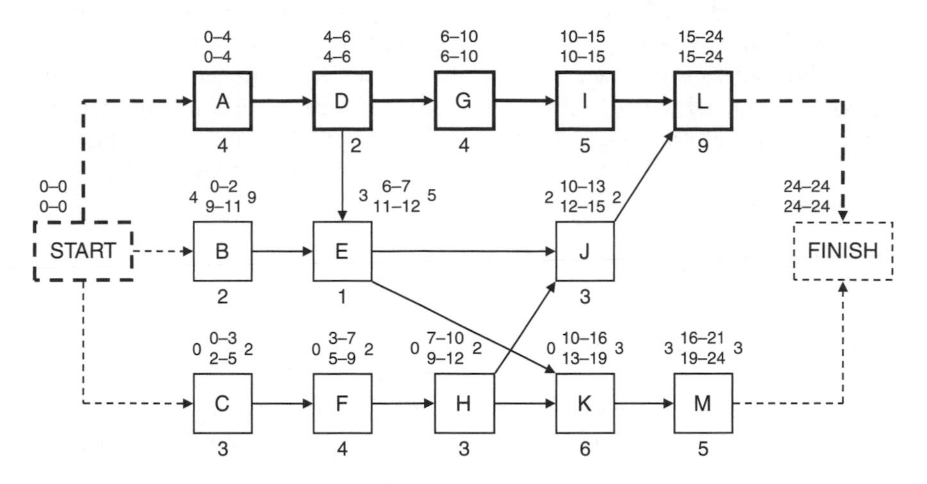

Figure 3.7 The calculated precedence schedule.

3.5.2 Free and total float

A bulk of the activities in most CPM schedules are non-critical. The notable feature of non-critical activities is that they can be delayed by a certain amount of time without delaying the earliest finish date of the schedule. While it is important to know which activities in a schedule are critical, it is equally important to know to what extent non-critical activities could be delayed without extending the scheduled completion date. This information is vital in managing the use of resources such as plant, equipment and labour. The topic of resource management will be discussed in detail in Chapter 4.

The amount of time by which a non-critical activity may be delayed is referred to as 'float'. When a non-critical activity has float, the planner may decide to start it between its ESD and LSD dates. As will be seen in Chapter 4, the decision on how much float is to be expended is commonly influenced by the availability of resources and the extent of delays that need to be minimised.

Four different float values can be calculated for non-critical activities. They are:

- Free float
- Total float
- Interfering float
- Independent float.

Only free and total float will be discussed in this book. Information on interfering and independent float can be found in Harris (1978).

'Free float' is defined as the amount of time by which a particular non-critical activity, I, in a schedule may be delayed without delaying the ESD of the succeeding activity, J. When activity I is followed by two or more succeeding activities, $J_1, J_2 \ldots J_n$, a free float value will be calculated for each link that activity I has with the following activities, $J_1, J_2 \ldots J_n$. The smallest of the individual free float values will become a free float of activity I. The general formula for calculating free float is:

$$\text{Free float}_I = \text{MIN}\,(\text{ESD}_{J_1, J_2 \ldots Jn} - \text{EFD}_I)$$

'Total float' is the amount of time by which a particular non-critical activity, I, in the schedule may be delayed without delaying the LSD of the succeeding activity, J. When an activity is scheduled to start at its LSD event, that activity is critical. If an LSD event of that activity is delayed, the completion date of the schedule will be delayed.

When the preceding activity, I, is followed by two or more succeeding activities, $J_1, J_2 \ldots J_n$, a total float value will be calculated for each link that activity I has with the succeeding activities, $J_1, J_2 \ldots J_n$. The smallest of the individual total float values will be a total float of activity I. The general formula for calculating total float is:

$$\text{Total float}_I = \text{MIN}\,(\text{LSD}_{J_1, J_2 \ldots Jn} - \text{EFD}_I)$$

The graphic definition of float in a precedence network is given in Figure 3.8 in regard to activities G and K. When the relationship between these two activities is illustrated in a bar chart format, the meaning of free and total float becomes immediately apparent.

It is obvious from the definition of free and total float that for any given non-critical activity I, its free float cannot be greater than its total float:

$$\text{Free Float}_I \leq \text{Total Float}_I$$

Free and total float of the non-critical activities B, C, E, F, H, J, K and M in Figure 3.7 will now be calculated. Let's start with activity B.

$$FF_B = ESD_E - EFD_B = 6 - 2 = 4$$

$$TF_B = LSD_E - EFD_B = 11 - 2 = 9$$

Free and total float of non-critical activity C is:

$$FF_C = ESD_F - EFD_C = 3 - 3 = 0$$

$$TF_C = LSD_F - EFD_C = 5 - 3 = 2$$

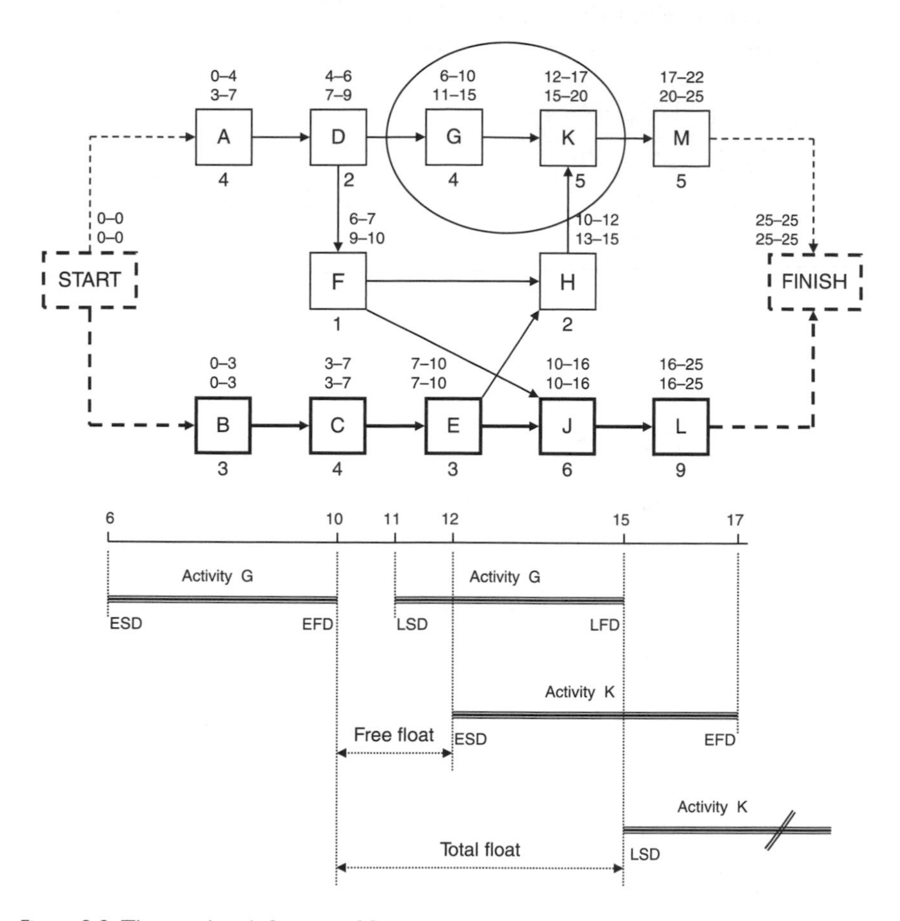

Figure 3.8 The graphic definition of float in a precedence network.

Non-critical activity E has two succeeding links with activities J and K. Therefore, free and total float will need to be calculated for each of these links. The minimum value of the individual free and total float values will then become a free float and a total float of activity E.

$$FF_E = ESD_J - EFD_E = 10 - 7 = 3 \text{ for link EJ and}$$

$$FF_E = ESD_K - EFD_E = 10 - 7 = 3 \text{ for link EK}$$

Since both individual free float values are the same, $FF_E = 3$.

$$TF_E = LSD_J - EFD_E = 12 - 7 = 5 \text{ for link EJ and}$$

$$TF_E = LSD_K - EFD_E = 13 - 7 = 6 \text{ for link EK}$$

Because total float of link EJ is smaller, $TF_E = 5$.

The remaining free and total float values are calculated as follows:

$$FF_F = ESD_H - EFD_F = 7 - 7 = 0$$
$$TF_F = LSD_H - EFD_F = 9 - 7 = 2$$
$$FF_H = ESD_J - EFD_H = 10 - 10 = 0 \text{ (for link HJ) and}$$
$$FF_H = ESD_K - EFD_H = 10 - 10 = 0 \text{ (for link HK), therefore, } FF_H = 0$$
$$TF_H = LSD_J - EFD_H = 12 - 10 = 2 \text{ (for link HJ) and}$$
$$TF_H = LSD_K - EFD_H = 13 - 10 = 3 \text{ (for link HK), therefore, } TF_H = 2$$

$$FF_J = ESD_L - EFD_J = 15 - 13 = 2$$
$$TF_J = LSD_L - EFD_J = 15 - 13 = 2$$

$$FF_K = ESD_M - EFD_K = 16 - 16 = 0$$
$$TF_K = LSD_M - EFD_K = 19 - 16 = 3$$

$$FF_M = ESD_{FINISH} - EFD_M = 24 - 21 = 3$$
$$TF_M = LSD_{FINISH} - EFD_M = 24 - 21 = 3$$

A fully calculated precedence schedule is given in Figure 3.7. It shows a critical path formed by activities A, D, G, I and L. It also shows the values of free and total float of the non-critical activities. It is interesting to note that some non-critical activities such as J and M have identical free and total float values, while the other non-critical activities do not. The reason is that the non-critical activities J and M have a succeeding link with the critical activities L and Finish respectively. Because critical activities have identical values of ESD and LSD, free and total float of preceding non-critical activities must therefore be identical.

3.5.3 Significance of float

Float may be viewed as extra time available in a schedule. If a CPM schedule has no float, it would have no capacity to accommodate delays. Apart from reducing the risk of time overruns, float also gives the planner an opportunity to manage allocated resources better.

From the scheduling point of view, float may be viewed as a time contingency. It gives the planner flexibility to schedule the start of a particular activity or activities within the ESD and LSD times.

Equally important is the contribution of float to effective management of committed resources. If all activities in a CPM schedule are required to start at their ESD times, it is unlikely that committed resources in such a schedule would be efficiently utilised. Float enables some non-critical activities to be delayed within the float limits to improve the efficiency of committed resource. For example, assume that two concurrent activities, one critical and the other non-critical, compete for the same resource, say a mobile crane. Clearly, these two activities cannot be performed concurrently with only one crane available. To resolve the problem, the planner may consider one of the following alternative solutions:

1 Assign the crane to the critical activity first and delay the start of the non-critical activity until the crane becomes available, provided the extent of the delay can be accommodated within the float limits.
2 Same as in 1, but the amount of float is insufficient to offset the delay. In this case the project completion date will be delayed.
3 Hire another crane at an extra cost to perform both activities concurrently. This alternative is conditional on availability of the second crane.
4 Reschedule the project.

The first alternative, which uses the available float, appears to offer the best solution, though the loss of float may increase the risk of future delays. The other solutions are likely to incur either extra cost or extra time or both. This and other issues concerning resource management will be discussed in more detail in the next chapter.

Since the presence of float in a CPM schedule is important for containing delays and achieving better efficiency of committed resources, who actually owns float needs to be fully understood. From the legal perspective, a party that is contractually bound to execute the work under the contract generally owns the float. For example, the contractor who under the terms of the contract is required to build a construction project owns the float during the construction period. If the client issues a variation order that causes a loss of float but does not delay the completion of the project, the contractor has no grounds to claim for extra time since the project has not been delayed. If, however, the amount of float is insufficient to offset the delay caused by the variation order, the contractor would have a legitimate claim for a time extension.

The loss of float caused by the client's variation order may also require the contractor to reallocate committed resources or increase the volume of resources in order to complete the work on schedule. In such cases, the contractor may be able to claim for cost if the contractor has prepared a resource-based schedule as a contract document (see section 1.5.4 in Chapter 1).

3.6 The concept of link lag

The mathematical procedure employed in critical path scheduling consists of forward and backward passes through the network. Networks with a small number of activities, such as the one in Figure 3.9, can easily be computed manually, but the complexity of computations increases with an increase in the size of the network. Computers are essential for scheduling large networks. When the first CPM software for use on personal computers was developed in the 1980s, these computers did not have the necessary computation power to handle the volume of information generated through the forward and backward passes. Simplification of the computational process was required, particularly of the backward pass, and some degree of inaccuracy of computer-generated CPM scheduling resulted.

The concept of 'link lags' offers a more efficient method of calculating a precedence network. It eliminates the backward pass calculations entirely. Apart from computing the forward pass, the remaining calculations are governed by simple formulas.

3.6.1 What is a link lag?

In a precedence network, a dependency line connecting a pair of preceding and succeeding activities is also referred to as a 'link lag' or simply a 'lag'. In reference to Figure 3.9, a link lag between activities I and J implies that the start of activity J lags after the finish of the preceding activity, I.

If activity J starts immediately after the completion of activity I, the lag value is zero. If, however, activity J starts some time after activity I has been completed, then the lag value will be positive. It is therefore possible to express lag using the following formula:

$$LAG_{IJ} = ESD_J - EFD_I$$

A closer examination of the above formula shows that LAG_{IJ} is expressed in the same manner as the free float of activity I. Therefore the free float of activity I is LAG_{IJ}. For more than one succeeding J activity, the general free float formula is:

$$\text{Free float}_I = \text{MIN}\ (LAG_{I\,J_1, J_2 \ldots J_n})$$

Total float can also be expressed using the concept of lag (Harris 1978). It can be derived from the expression of total float given in section 3.5.2 and from other expressions of total float that can be derived from Figure 3.8. Three formulas of total float have been derived. They are as follows:

$$TF_I = LSD_J - EFD_I \qquad (1)$$

or

$$TF_I = LFD_I - EFD_I \qquad (2)$$

or

$$TF_I = LSD_I - ESD_I \qquad (3)$$

If the total float formula (1) expressed for activity I holds, it must also hold for

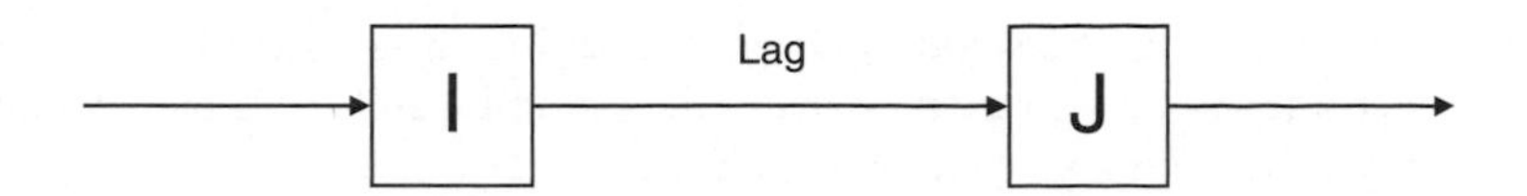

Figure 3.9 A link lag in the precedence network.

activity J or any other activity. Hence,

$$TF_J = LSD_J - ESD_J \tag{4}$$

and by rearranging the formula,

$$LSD_J = TF_J + ESD_J \tag{5}$$

Substituting the above expression (5) of LSD_J to the formula (1) results in:

$$TF_I = TF_J + ESD_J - EFD_I$$

and by rearranging the formula,

$$TF_I = (ESD_J - EFD_I) + TF_J$$

The expression in brackets ($ESD_J - EFD_I$) is in fact LAG_{IJ}. Therefore,

$$TF_I = LAG_{IJ} + TF_J$$

With more succeeding activities J_1 J_2 . . . J_n, the general formula for total float is:

$$\text{Total float}_I = \text{MIN}\ (LAG_{I\,J_1, J_2 \ldots J_n} + TF_J)$$

It is worth noting that in order to calculate total float of the preceding activity, I, the planner needs to know the value of total float of the succeeding activity, J. Since critical activities have zero total float, calculation of total float of preceding non-critical activities will start from succeeding critical activities, one of which will be the terminal activity that gives the total schedule duration.

3.6.2 Link lag process

The computation process of a precedence schedule using a link lag approach involves a number of sequential steps:

Step 1: Perform a forward pass and calculate the ESDs and EFDs of all activities in a schedule.
Step 2: Calculate link lags.
Step 3: Identify a critical path. The critical path is a path of zero lags.

Step 4: Calculate free and total float.

Step 5: Calculate the LSDs and LFDs of all activities in a schedule. Instead of performing a backward path, the values of LSDs and LFDs are calculated from the total float formula (2) above:

$$TF_I = LFD_I - EFD_I$$

From this expression, the value of the LFD is calculated as:

$$LFD_I = TF_I + EFD_I$$

The activity's LSD is then calculated by deducting its duration from the value of the LFD.

Let's now calculate a simple precedence schedule using the concept of lag. A schedule in given in Figure 3.10.

Step 1 involves a forward pass. Values of ESDs and EFDs of the activities in the schedule have already been calculated. In Step 2 lags are calculated. A list of all lags in the schedule is given below together with individual calculations:

- $Lag_{AB} = ESD_B - EFD_A = 4 - 4 = 0$
- $Lag_{AC} = ESD_C - EFD_A = 4 - 4 = 0$
- $Lag_{AD} = ESD_D - EFD_A = 4 - 4 = 0$
- $Lag_{BE} = ESD_E - EFD_B = 12 - 12 = 0$
- $Lag_{CE} = ESD_E - EFD_C = 12 - 7 = 5$
- $Lag_{CF} = ESD_F - EFD_C = 7 - 7 = 0$
- $Lag_{DE} = ESD_E - EFD_D = 12 - 6 = 6$

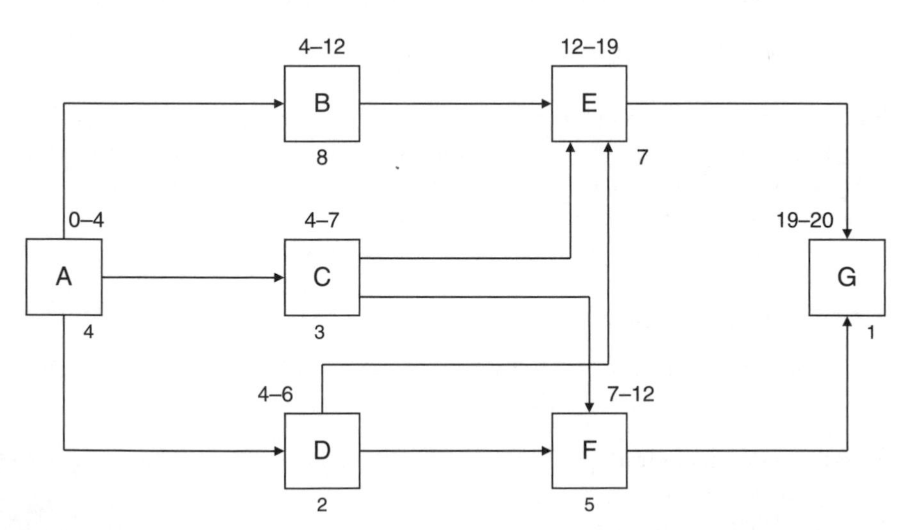

Figure 3.10 Example of a precedence schedule.

- $Lag_{DF} = ESD_F - EFD_D = 7 - 6 = 1$
- $Lag_{EG} = ESD_G - EFD_E = 19 - 19 = 0$
- $Lag_{FG} = ESD_G - EFD_F = 19 - 12 = 7$

The above lag values have been added to a schedule in Figure 3.11. It should be noted that the values of lags are also values of free float. Step 3 requires identification of a critical path. Since it is defined as a path of zero lags, it is then a simple task to locate a critical path. There is only one path of zero lags in a schedule in Figure 3.11. It connects the critical activities A–B–E–G. In Step 4, free and total float are calculated. Since critical activities have no float, the calculation process will only involve non-critical activities. Free float have already been calculated as lags and it is therefore a simple task of assigning them to each non-critical activity. Consequently,

$FF_C = MIN\ (LAG_{CE\ CF})$ where $LAG_{CE} = 5$ and $LAG_{CF} = 0$, therefore $FF_C = 0$

$FF_D = MIN\ (LAG_{DE\ DF})$ where $LAG_{DE} = 6$ and $LAG_{DF} = 1$, therefore $FF_D = 1$

$FF_F = LAG_{FG} = 7$

Total float, however, needs to be calculated from the formula derived previously. That formula requires the value of total float of the following activity, J, to be known. The last activity, G, in the schedule in Figure 3.11 is critical. Its total float is zero. Therefore, it is now possible to calculate total float of the preceding non-critical activity, F, as follows:

$TF_F = LAG_{FG} + TF_G = 7 + 0 = 7$

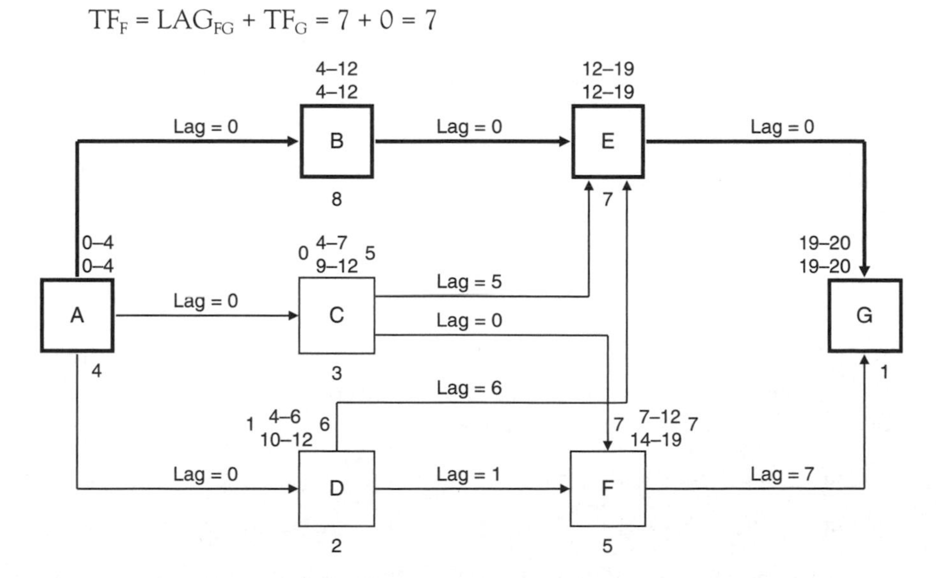

Figure 3.11 Example of a precedence network including lags and the critical path.

Having previously calculated total float of the non-critical activity F, and given that total float of the critical activity E is zero, it is now possible to calculate total float of the non-critical activity C, which has two succeeding links, one with the critical activity E and the other with the non-critical activity F. Consequently,

$$TF_C = LAG_{CE} + TF_E = 5 + 0 = 5$$

and

$$TF_C = LAG_{CF} + TF_F = 0 + 7 = 7$$

Therefore,

$$TF_C = 5$$

Similarly,

$$TF_D = LAG_{DE} + TF_E = 6 + 0 = 6$$

and

$$TF_D = LAG_{DF} + TF_F = 1 + 7 = 8$$

Therefore,

$$TF_D = 6$$

The values of free and total float are given in Figure 3.11.

In the last step, values of LSD and LFD of all the activities in the schedule are calculated from a simple total float formula:

$$TF_I = LFD_I - EFDI_I$$

Then

$$LFD_I = TF_I + EFD_I$$

The LSD and LFD values of the critical activities are already known since they are the same as the ESD and EFD values respectively. Let's now calculate the LSD and LFD values of the non-critical activities.

$$LFD_C = TF_C + EFD_C = 5 + 7 = 12,\ LSD_C = LFD_C - duration_C = 12 - 3 = 9$$

$$LFD_D = TF_D + EFD_D = 6 + 6 = 12,\ LSD_D = LFD_D - duration_D = 12 - 2 = 10$$

$$LFD = TF + EFD = 7 + 12 = 19,\ LSD_F = LFD_F - duration_F = 19 - 5 = 14$$

The entire computation process has now been completed and all the necessary information has been derived. The fully calculated schedule is given in Figure 3.11.

3.7 Summary

This chapter has presented the basic concept of critical path scheduling. Initially, the main features of the now largely superseded arrow method were presented followed by a detailed examination of the precedence method of critical path scheduling including network construction, forward and backward path calculation, identification of a critical path, and calculation of float. In the final part of this chapter the concept of link lag as an alternative method of critical path scheduling was defined and its computational simplicity demonstrated on a simple example.

The concept of critical path scheduling will be further expanded in other chapters of this book. The next three chapters will discuss applications of the CPM method in resource scheduling, overlapping of activities, and in monitoring and controlling of projects respectively. Computer-based CPM scheduling will be discussed in detail in Chapter 7. The extension of the CPM to critical chain scheduling will then be explored in Chapter 8, and probability scheduling using the Monte Carlo simulation and PERT techniques will be presented in Chapters 12 and 13 respectively.

Exercises

Solutions to the following exercises can be found on the following website: http://www.routledge.com/books/details/9780415601696/

Exercise 3.1

Plot a precedence schedule for the list of activities shown in Table 3.2.

Table 3.2 Data for a precedence schedule

Activity	*Depends on completion of these activities*
C	A
D	A
E	A
F	D
G	C and D
H	E
J	B
K	B and E
L	B, D and E
M	F and G
N	H, K and L
O	G and J
P	M, N and O

Exercise 3.2

Plot a precedence schedule for the list of activities shown in Table 3.3.

Table 3.3 Data for a precedence schedule

Activity	*Depends on completion of these activities*
C	A
D	A
E	B
F	B
G	C
H	B and D
J	B and D
K	J
L	J
M	E
N	F
O	E, G, H, K and L
P	M and N
Q	M and O

Exercise 3.3

Plot a precedence schedule for the list of activities shown in Table 3.4.

Table 3.4 Data for a precedence schedule

Activity	*Depends on completion of these activities*
B and C	A
G and K	F
H	E and G
D and J	B, C and H
L	J and K
N	D, J and K
M	L

Exercise 3.4

A list of activities forming a precedence schedule is given in Table 3.5.

Table 3.5 Data for a precedence schedule

Activity	*Depends on completion of these activities*	*Duration*
A	START	2
B	START	3
C	START	4

D	A and B	4
E	B	10
F	B	12
G	C	6
H	A	5
J	D, E and H	12
K	F and G	4
L	G	7
M	L	7
N	J	4
O	N	3
P	J and L	5
Q	J, K and M	6
R	N and P	3
S	O	8
T	Q	10
U	Q and R	5

a) Plot a precedence schedule.
b) Calculate the schedule using the forward and the backward path method. Determine the ESD, EFD, LSD and LFD values for all the schedule activities.
c) Determine the position of a critical path.
d) Calculate free and total float of all non-critical activities.
e) Fully recalculate the schedule using the link lag approach.

Exercise 3.5

A list of activities forming a precedence schedule is given in Table 3.6.

Table 3.6 Data for a precedence schedule

Activity	*Depends on completion of these activities*	*Duration*
A	START	5
B	START	3
C	A	3
D	A	2
E	B	2
F	B	4
G	C	7
H	B and D	2
J	B and D	4
K	J	6
L	J	5
M	E	7
N	F	6
O	E, G, H, K and L	5
P	M and N	11
Q	M and O	3

a) Plot a precedence schedule.
b) Calculate the schedule using the forward and the backward path method. Determine the ESD, EFD, LSD and LFD values for all the schedule activities.
c) Determine the position of a critical path.
d) Calculate free and total float of all non-critical activities.
e) Fully recalculate the schedule using the link lag approach.

Exercise 3.6

Apart from having a link with the succeeding activity B, activity A has no further succeeding links. Duration of activity A is four days. Scheduling information for activity A is given below.

ESD_A	4	FF_A	5
EFD_A	8	TF_A	7

Activity B has a link with the preceding activity A and may have links with other unspecified preceding activities. Duration of activity B is two days.

Calculate the following information:

a) ESD_B ......
   EFD_B ......
b) LSD_B ......
   LFD_B ......
c) LSD_A ......
   LFD_A ......

Chapter 4

Resource management

4.1 Introduction

Managing resources is vital in planning construction projects. Specifically, this chapter looks at how resources are distributed throughout a project. It also examines the process of resource levelling, and the management of labour, material and plant/equipment resources.

The concept of critical path scheduling was described in the previous chapter. The discussion has thus far been restricted to the construction of networks in the form of time schedules. Time schedules were defined in Chapter 1 as plans that are concerned with developing the overall production strategy within a given time-frame. They tend to be overly optimistic because they largely ignore resources. Time scheduling in fact assumes that resources are unlimited and can be allocated to activities in a schedule whenever needed. Despite ignoring resources, they are nevertheless useful in long-range planning when all that a planner is interested to know is the total time-frame for a project and the overall production strategy.

Time schedules are, however, inappropriate in medium- to short-range scheduling where efficiency in the use of resources is of the utmost importance. There are two reasons for this. First, the availability of required resources is almost always limited in some way, for example in quantity of resources or in the required level of skill and technical specification. The use of resources may also be limited by their cost. To illustrate this, let's assume that according to a time schedule the contractor is required to repair the sandstone façade of a historical building within four months. To meet the scheduled completion time, the contractor expects to employ eight highly skilled stonemasons. But only six suitably qualified persons are available locally for work. The contractor's options are: to bring additional stonemasons from interstate or overseas at an extra cost; to do the work with six stonemasons only and risk delaying the contract with the consequence of incurring liquidated damages; or to try to renegotiate the contract with the client. The assumption of 'unlimited resources' in the time schedule has clearly placed the contractor in a difficult position.

The second reason is that time scheduling ignores the efficiency of committed resources. Let's assume that a time schedule for construction of a high-rise building requires a number of activities such as formwork, reinforcement, air conditioning ductwork, concreting, brickwork, precast concrete façade, scaffolding, etc., all to be performed concurrently within specific periods. Let's assume further that the contractor has committed to the project a specific type of tower crane based on the contractor's expectation that it can handle the total volume of work to be performed. The crane has already been erected on the site. While the crane may have the capacity to handle the total volume of work, the contractor learns rather too late that it is unable to handle the daily volume of work within specific periods when a number of activities are scheduled concurrently. If the contractor is unable to increase the capacity of the crane or supplement it with additional hoisting equipment, delays in the execution of the work are likely. To minimise such a possibility, the contractor would attempt to reschedule the work around the committed crane by using the available float (spare time) in non-critical activities. This will require delaying the start of some of non-critical activities in an attempt to reduce the resource demand peak. However, the amount of float may not always be sufficient to permit effective rescheduling of the work around committed resources.

Resource scheduling overcomes shortcomings of time scheduling by assuming that resources are always limited in some way. Its function is to allocate work efficiently to what resources have already been committed. By considering resources, a schedule becomes more realistic and more representative of the actual production process.

The critical path method (CPM) is suitable for resource scheduling. The actual process of resource scheduling involves first constructing a time schedule and then converting it to a resource schedule through a process of resource levelling. Other scheduling techniques such as a 'multiple activity chart' and a 'line of balance' are effective in resource scheduling and are discussed in detail in Chapters 9 and 10.

4.2 Resources

According to Wikipedia, 'a resource is any physical or virtual entity of limited availability that needs to be consumed to obtain a benefit from it'. Examples of resources are:

- Time
- Labour
- Plant or equipment
- Materials
- Money.

Time is the main resource in time scheduling, while resource scheduling is concerned with the most efficient use of labour, plant/equipment and materials. Managing labour, plant/equipment and materials will be examined later in this chapter. Money is obviously an important resource in construction, but its management lies outside the scope of this book.

4.3 Distribution of resources

The financial success of a project is largely dependent on the ability of a project manager to employ resources efficiently. If resources are employed inefficiently, for example a committed crane only works 60 per cent of the time or a group of carpenters is idle for many hours because there is no work for them, a project incurs extra costs. The more inefficient the project is in its use of resources, the more extra costs it will incur. It is a challenge for a project manager to ensure that resources are used to their maximum efficiency, thus keeping the cost down. This requires:

- An orderly and even flow of work
- Continuous work without interruptions (idle time costs money)
- An adequate volume of allocated resources
- The employment of appropriately skilled labour resources, technologically adequate plant/equipment resources and material resources of the highest quality
- The employment of a correct mix of labour to plant/equipment.

4.3.1 Even or uniform distribution of resources

Efficiency in the use of resources is achieved when they are distributed evenly as illustrated in Figure 4.1. But it would be unrealistic to expect to achieve such an even distribution over the entire period of the project.

Uneven distribution of resources on construction projects is largely related to the fluctuating volume and intensity of work over the project period. In the initial construction period, involving mainly ground works, only a small number of resources, both human and physical, are engaged. Thereafter, more and more subcontractors and other resources join the project until at the peak of

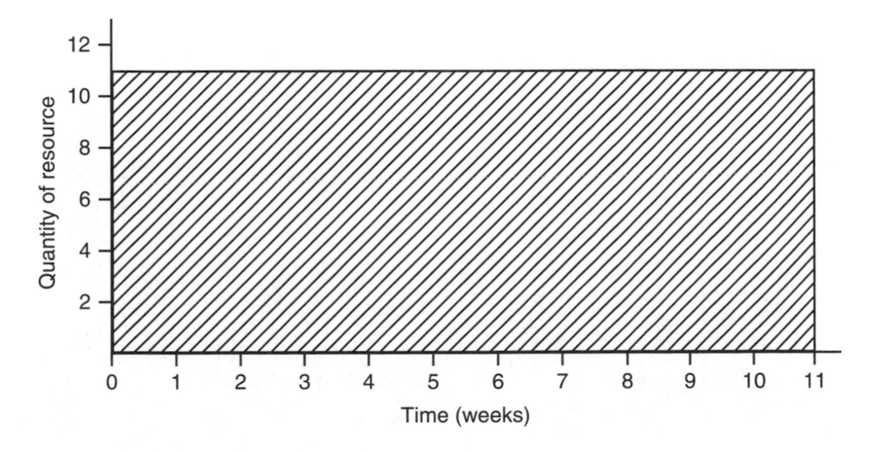

Figure 4.1 Example of the uniform distribution of a resource.

construction activity, which usually occurs about two-thirds into the project period, all the resources are engaged. In the final phase, individual subcontractors and other resources gradually withdraw from the project when they are no longer needed. The volume and intensity of work then gradually diminishes until the project is fully accomplished at the end of the contract period.

4.3.2 Uneven distribution of resources

The labour resource tends to be unevenly distributed over the period of a construction project. Let's assume that Figure 4.2 illustrates distribution of the labour resource of a civil engineering contractor engaged in construction of some drainage work. Assume further that the contractor's workers are multi-skilled and capable of performing all the activities associated with this project. The scheduled resource demand in Figure 4.2 shows a highly inefficient use of the labour resource. It is unlikely that the contractor would attempt to build the project by varying the number of workers from day to day to meet the scheduled resource demand in Figure 4.2. Instead, the prudent contractor would attempt to reduce the demand peaks and minimise the demand troughs by applying the concept of 'levelling', which will be discussed later in this chapter.

4.3.3 Normal, skewed and complex distributions of resources

Between the extremes of even and uneven resource distributions, there are many other distributions of resources of which normal, skewed and complex are worth noting. The normal and in particular the right-skewed distribution are examples of ideal distributions of resources on construction projects. This is because they take into account a low level of activity in the beginning and the tapering off of

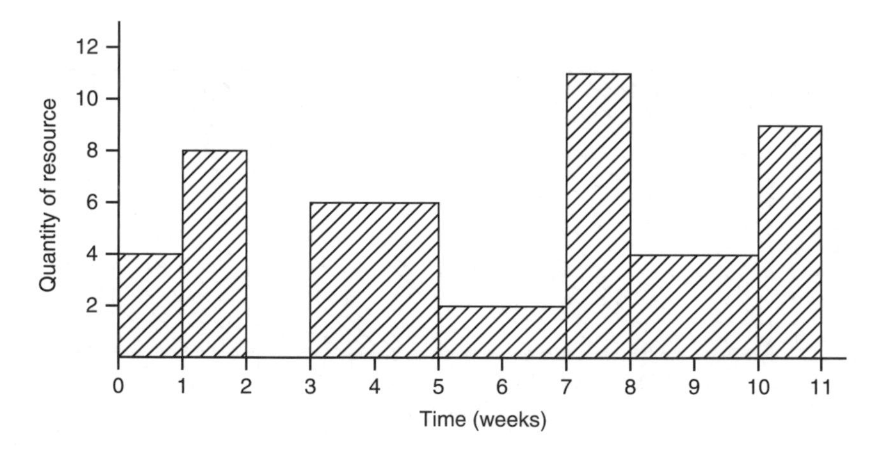

Figure 4.2 Example of the uneven distribution of a resource.

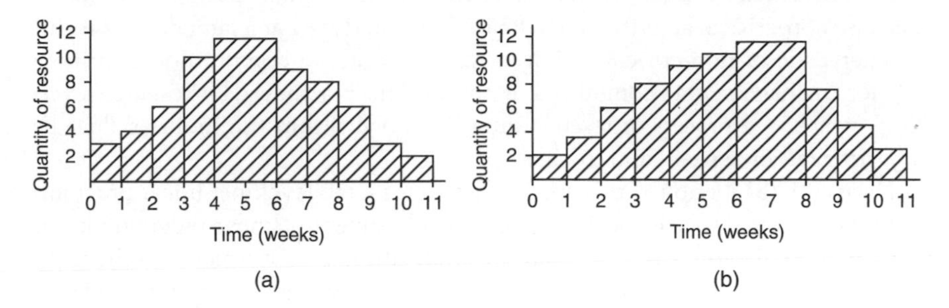

Figure 4.3 Example of the normal and the right-skewed distributed resources.

the work at the end of the project. These distributions are shown graphically in Figure 4.3(a) and (b) respectively.

Managing one unevenly distributed resource may be difficult enough, but managing an array of unevenly distributed resources such as those given in Figure 4.4, which is characteristic of most construction projects, is an extremely challenging task. In CPM software, however, project managers have a powerful tool for managing even most complex distributions of resources. A brief overview of how computer software can manage resources will be given in Chapter 7.

4.4 Resource levelling

The previous section has described different patterns of resource distribution and highlighted those that are desirable for achieving resource efficiency. Since the distribution of resources in construction projects is largely uneven, the task of the planner is to ensure their best possible use.

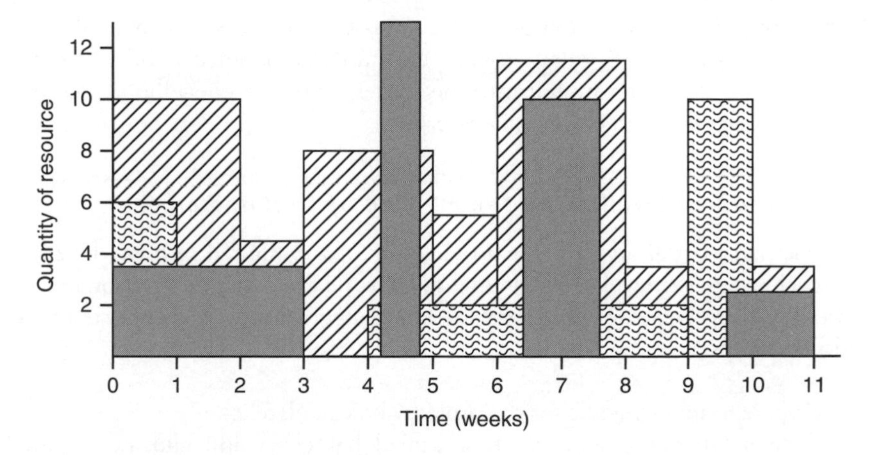

Figure 4.4 A complex distribution of resources.

A time schedule requires that activities begin at their earliest start dates. Because non-critical activities have float, they could start at a later date. It is this characteristic of non-critical activities that provides an opportunity for a more efficient management of committed resources. This is achieved by 'resource levelling', which is a process of rescheduling the work within the limits of float by adjusting resource peaks and troughs.

Harris (1978) identified two approaches to resource levelling: linear programming and heuristics. Although mathematically superior, linear programming is computationally intensive, particularly in relation to construction schedules. Heuristic processes offer low computation intensity but may not provide an optimum solution. It was because of computational intensity that linear programming came to be passed over in favour of heuristic processes. Harris (1978: 255) defines heuristics as 'a set of rules of thumb designed to progressively lead the user [by trial and error] to a feasible solution'.

The development of heuristic processes has followed two distinct paths: unlimited resource levelling and limited resource levelling. Unlimited resource levelling attempts to minimise the resource input and therefore its cost while maintaining the project duration generated by the CPM schedule. Limited resource levelling attempts to minimise the project duration while keeping the resource levels constant.

In the practical sense, since the finish date of construction projects is commonly fixed by a contract, unlimited resource levelling is a more appropriate method to use. The most commonly used unlimited method is the sum of squares of the daily resource demands. Most CPM software performs resource levelling using this method. An alternative approach, referred to as the 'minimum moment method', was proposed by Harris (1978). It is based on the sum of squares method with the addition of an 'improvement factor' that selects the activity to be reallocated along its float.

Manual resource levelling of only one resource is extremely time-consuming. Considering that construction projects comprise many resources, manual levelling is just not feasible. Computers can perform this task quickly and efficiently. However, to get a better insight into the process of resource levelling, a method of trial and error will now be briefly described.

4.4.1 Resource levelling by the method of trial and error

The method of trial and error is an example of the limited resource-levelling approach where the level of resources is fixed while the schedule duration may be varied. It represents the most basic heuristic algorithm, which consists of the following steps:

1 Prepare a time schedule for the project and calculate it.
2 Convert the time schedule to a scaled bar chart and allocate required resources to each activity.

3 Calculate total daily resource sums for each day of the schedule.
4 Plot a histogram of the resource demand (one for each resource).
5 Determine resource availability on a day-to-day basis. Compare demand and availability, and determine whether the demand histogram is acceptable.
6 Evaluate possible alternatives for levelling using the resource utilisation factor (RUF).

$$\text{RUF} = \frac{\text{Usable resource} \times \text{Days available}}{\text{Usable resource} \times \text{Days used}} \times 100\%$$

7 Implement levelling by rescheduling selected activities within the limits of their float.

The levelling method of trial and error attempts to move non-critical activities along their float away from peak periods of resource demand while keeping the resource input levels constant. The aim is to achieve best possible use of committed resources without unduly extending the project period. In shifting non-critical activities along their float, care is needed to ensure that the logical links between the activities are maintained.

4.4.2 Example of resource levelling using the method of trial and error

Resource levelling will be performed on a simple precedence schedule in Figure 4.5 for one resource only. Assume that the resource in question is labour. Let's further assume that the workers are multi-skilled and therefore able to perform any activity in the schedule. The maximum number of available workers

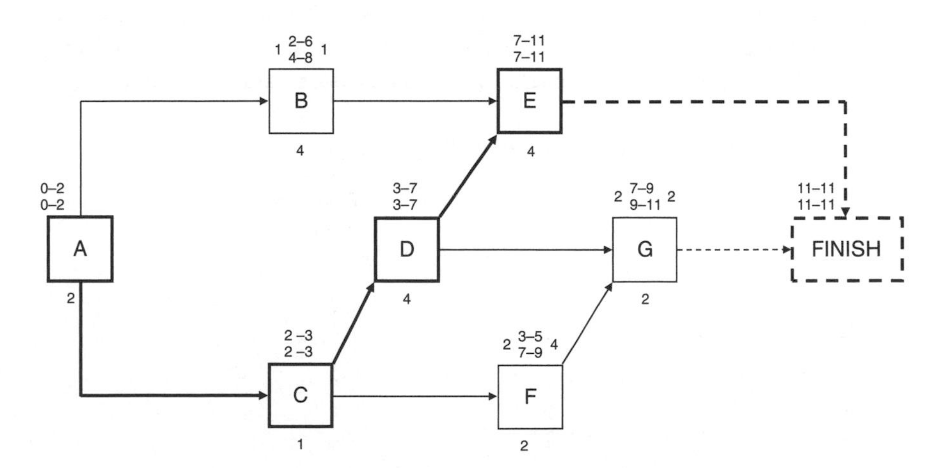

Figure 4.5 A precedence network for the resource-levelling example.

for this job is seven. The contractor's task is to ensure that this project could be accomplished with seven workers only.

The resource data for this project are given in Table 4.1.

Resource levelling using the method of trial and error is best performed on a scaled bar chart. Let's begin resource levelling step by step.

Step 1. Prepare a time schedule for the project and fully calculate it
The fully calculated precedence schedule is given in Figure 4.5.

Step 2. Convert the time schedule to a scaled bar chart and allocate resources to each activity
A time schedule in Figure 4.5 was converted to a linked bar chart (see Figure 4.6). The bar chart is organised so that the critical activities are separated from the non-critical ones. Each activity displays its daily labour resource rate.

Step 3. Calculate total daily resource sums
At the base of the bar chart, the total daily resource sums are calculated. The total daily resource sums $\Sigma 0$ reflect the distribution of the labour resource for the original time schedule, which ranges from two to 11 workers.

Step 4. Plot a histogram of the labour resource demand
The labour resource demand histogram is given in Figure 4.7(a). It shows the demand peak of 11 workers on day 3. With only seven workers available for this job, the demand levels on days 3, 4 and 5 cannot be met.

Step 5. Level the labour resource
The peak of the resource demand on day 3 was caused by two concurrent activities, B and C. Activity B has one day of total float. If moved by one day to the right, the peak of 11 persons on day 3 will be reduced to six and the resource daily sum on day 7 would increase from two to seven. It is therefore logical to commence levelling by moving activity B by one day. This is noted on the bar chart as 'B → 1'. The new total daily labour sums $\Sigma 1$ are then calculated (see Figure 4.6). It should be noted that activity B has now become critical since its float has been used up.

Table 4.1 Data for the resource-levelling example

Activity	*Duration (days)*	*Resource rate per day*	*Resource days*
A	2	3	6
B	4	5	20
C	1	6	6
D	4	2	8
E	4	4	16
F	2	2	4
G	2	2	4
Total			64

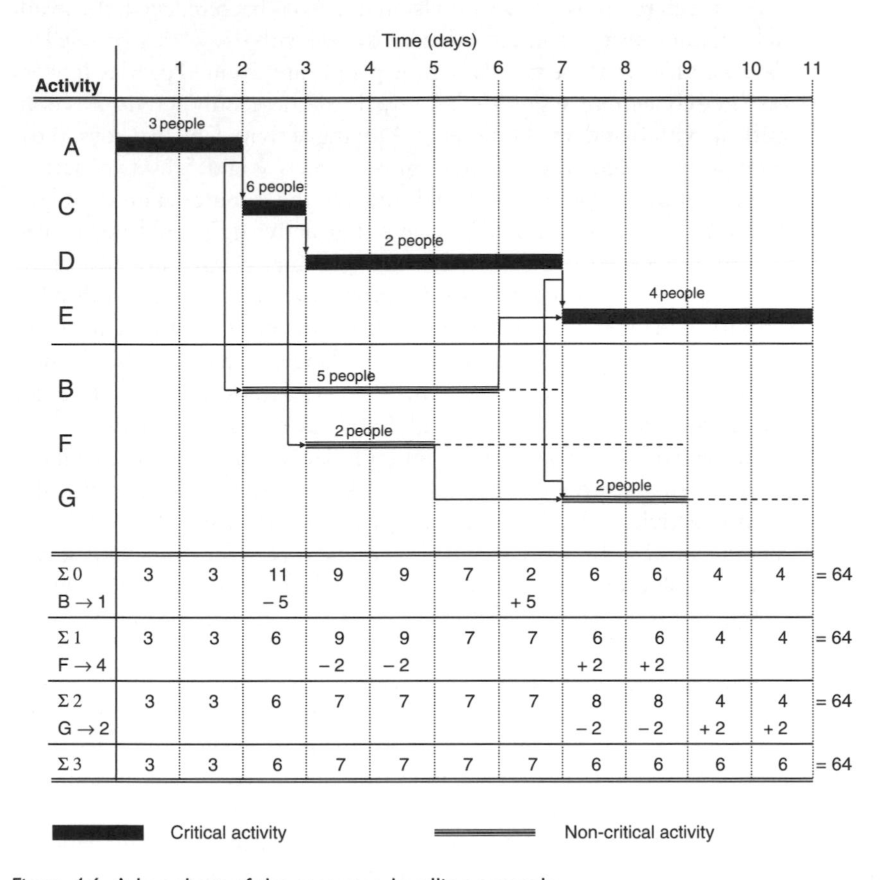

Figure 4.6 A bar chart of the resource-levelling example.

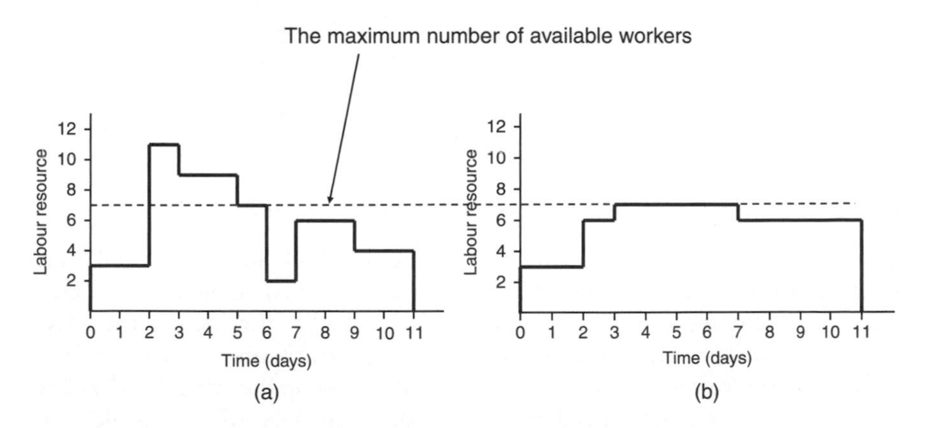

Figure 4.7 A histogram of the labour demand before and after levelling.

Although the peak of 11 workers on day 3 has been reduced, the available labour resource is still insufficient to deal with the work scheduled for days 4 and 5. On these two days, nine people are required per day to work on three concurrent activities, B, D and F. Of these, only activity F is non-critical, with four days of total float. Moving activity F by two days to the right would reduce the resource sums on days 4 and 5 by two days to seven, but at the same time would increase the resource sums on days 6 and 7 by two days to nine. It means that activity F would need to be moved by a further two days; this is possible since it has four days of total float. However, shifting activity F beyond two days requires shifting of activity G in order to maintain the logical order of work. When moved by four days, activity F will increase the daily labour demand on days 8 and 9 from six to eight. However, when activity G is moved by two days, the labour demand on days 8 and 9 will drop back to six. After activity F has been moved by four days, the total daily labour sum Σ2 was calculated within the range from three to eight workers. Activity F is now critical.

After activity G has been shifted by two days, the calculated total daily labour sums Σ3 show the range of workers as between three and seven. The available crew of workers is now able to meet the daily labour demand. It should be noted that activity G has also become critical.

Step 6. Evaluate the levelling process

The levelling process described above provides a solution for the given problem, that is, the project can be accomplished by employing the maximum of seven workers. The labour demand histogram after levelling given in Figure 4.7(b) shows that the labour resource is now better utilised. This is verified by the calculation of the RUF, which has improved from 52.9% to 83.1%. (Note that in calculating the denominator in the above equations it was assumed that the peak of the resource demand would be required for the duration of the project.)

$$\text{RUF (before levelling)} = \frac{64 \text{ persons days planned}}{11 \text{ (resource peak)} \times 11 \text{ days}} \times 100\% = 52.9\%$$

$$\text{RUF (after levelling)} = \frac{64 \text{ persons days planned}}{7 \text{ (resource peak)} \times 11 \text{ days}} \times 100\% = 83.1\%$$

4.3.3 The impact of resource levelling on a schedule

The process of resource levelling reallocates the work in the original time schedule to achieve the most efficient use of committed resources. The final outcome of resource levelling is a resource-based schedule in which all the activities are critical in terms of committed resources. Furthermore, there are also those activities that are critical in terms of time (those on the critical path). The transformation of a time schedule to a resource schedule, which was demonstrated on the above example, raises a number of interesting issues:

1 *Resource levelling reduces the amount of float*
In the above example, the total float of the non-critical activities was fully used up, turning the non-critical activities B, F and G into critical ones. With more critical activities and less float, the ability of the resource schedule to absorb future delays is severely reduced. When scheduling a project, the planner will need to consider carefully the trade-off between a more efficient use of resources on one hand and the loss of float on the other.

2 *Resource levelling shifts the work more towards the end of the project*
Examination of the labour demand histograms before and after levelling clearly shows that resource levelling shifts more work to the end of the project. The accumulation of more work at this stage places more emphasis on coordination of activities, for which more supervisors may be needed. Another implication is that the contractor's cash flow will be altered.

3 *Resource levelling locks activities into specific start and finish dates*
In resource levelling, non-critical activities are shifted along their float to specific start and finish dates. The failure to start those activities as scheduled will cause the level of resource demand to be out of balance with the level of committed resources. It may therefore be said that after a schedule has been resource-levelled, all of its activities are critical in terms of committed resources, even if they have float. They must start and finish as specified. It follows that a resource-levelled schedule may be difficult to maintain where the risk of potential time overruns is high. A prudent approach to resource scheduling is to ensure that precautionary time and resource buffers are built into the schedule to mitigate the impact of future unforeseen delays.

4.3.4 Resource levelling of critical activities

Some schedules may contain a large proportion of critical activities, or critical activities may be dominant in their specific parts. Levelling such schedules is obviously difficult in the absence of float. However, if the planner has some flexibility in varying the distribution of a resource day by day, two potential approaches to resource levelling could be adopted: to vary the distribution of a resource from day to day and/or splitting and varying the distribution of a resource. Let's examine these two approaches in turn.

The first approach is based on the assumption that the resource may not always be distributed evenly from day to day. For example, the activity 'painting' is performed by a crew of three workers over six days. The planner would initially allocate the labour resource evenly by committing three workers to each day of the painting activity. However, the planner may be able to vary the number of workers from day to day while keeping the total labour demand constant. For example, the planner may commit four painters per day for the first three days and two per day for the remaining three days.

This may or may not be possible. The production rate of some activities such as excavation, steelfixing, painting, electrical and the like may possibly be increased proportionally to the increase in the volume of the resource. However, the same may not hold true for other activities that may be constrained in their performance by limited space, strict safety issues or specific design requirements such as those related to curing of materials.

Let's illustrate the method of varying the distribution of a resource on a simple example in Figure 4.8(a). Activities A, B and C are critical. Assume that the distribution of the resource in activities A and B can be varied. Let's further assume that only seven resource units are available to perform these three critical activities.

Since the activities are critical, no shifting along float is possible. But the distribution of the resource in activities A and B can be altered to bring the peak demand to seven units. This is illustrated in Figure 4.8(b).

The second approach uses the ability of some activities to have specific periods designated as non-working. This has the effect of splitting an activity (it should be noted that not all activities are capable of being split). If the distribution of a resource can be varied within the split parts of the activity, the resource demand may be controlled. This approached is illustrated in Figure 4.9(a) and (b). After the resource has been split and varied in intensity, the scheduled work can be performed with seven resource units.

Resource levelling by varying or by splitting the distribution of the resource is not restricted to critical activities but may be applied in the same manner to non-critical activities.

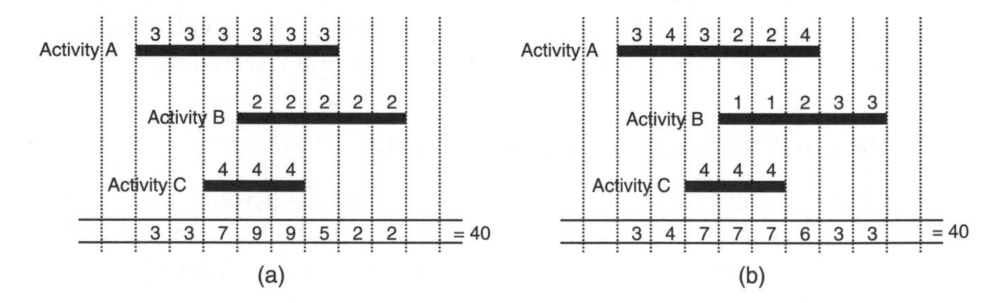

Figure 4.8 Lowering the resource demand by varying the distribution of the resource.

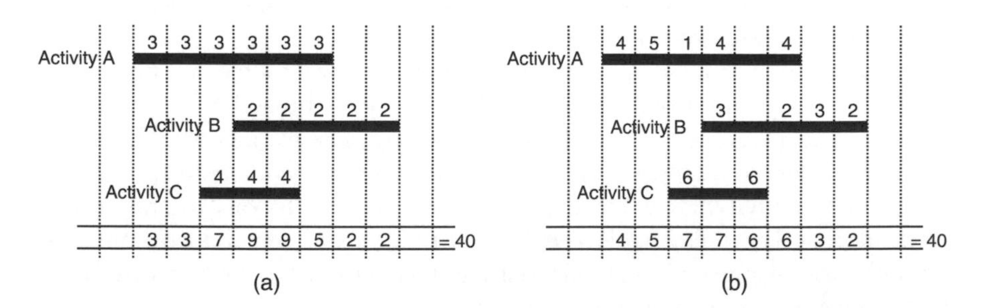

Figure 4.9 Lowering the resource demand by splitting and varying the distribution of the resource.

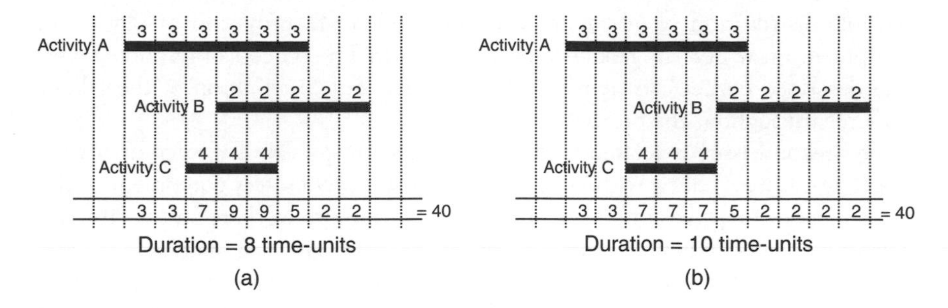

Figure 4.10 Lowering the resource demand by extending the project duration.

If the distribution of a resource cannot be varied or split and the activities have no float, extending the project period may be the only solution. Lowering the resource demand to seven units can achieved by delaying activity B, and hence the whole project, by two days (see Figure 4.10(a) and (b)).

4.5 Resource levelling performed by computers

The method of trial and error (heuristics) has been used to explain the mechanics of resource levelling on a simple example involving a single resource. Levelling of a multitude of resources is a highly complex problem that is best performed by computers.

Most CPM software relies on the concept of unlimited resource levelling, which ensures that the duration of the project is not extended. Resource-levelling algorithms built into such software are commonly based on the statistical concept known as 'the sum of the squares of the daily resource demands'. Some software is highly refined and most reliable in its ability to level resources effectively, while some is relatively crude and less accurate. The levelling algorism of popular Primavera P6 software will be briefly examined in Chapter 7.

4.6 Managing the labour resource

The construction industry is a large employer of labour. Some construction workers are employed directly by contractors, while most are employees of subcontractors. Irrespective of who employs construction workers and when they are engaged in construction activities the contractor is responsible for the management of the entire workforce. There are a number of reasons for this. The first and probably the most important reason is efficiency: the contractor needs to integrate the work of directly employed workers and subcontractors within the overall project planning strategy. This requires coordination of the activities of preceding and succeeding subcontractors to ensure continuity of work, as even as possible distribution of labour resources and adequate provision of plant and equipment on which the labour resource depends. The process of resource levelling described earlier can help in achieving efficient distribution of the labour resource throughout the project.

The second issue is related to the contractor's responsibility to comply with relevant statutes dealing with safety and the provision of safety equipment to the workers, and the provision of site amenities. With more workers on the site, the demand for safety equipment and amenities increases proportionally and so does the cost. If the contractor is able to decrease the labour resource from 150 workers to 100 without an undue increase in the time schedule, savings in the areas of safety and amenities are clearly apparent.

Having a very large number of workers engaged on a construction site can also create a problem in the efficient handling of personnel. Let's assume that a large high-rise commercial project will require at the peak of construction activities the services of around 2,000 workers if it is to be completed within the expected period. How will the contractor be able to effectively distribute such a large number of people through the job?

This is a very complex problem that requires, apart from determining the number of personnel hoists, consideration of a number of important issues. These may range from traffic congestion around the site created by the arrival and the departure of so many workers in the morning and the afternoon, congestion on the site itself created by the large workforce and the potential problems of managing so many people, to the loss of productivity caused by the time wasted while waiting for hoists. One possible solution would be to stagger the hours of work to reduce congestion. But in most urban areas, local development authorities may not permit such an extension outside the usual 7 am to 6 pm range. It may well be that due to restrictions imposed on the project by some of the factors mentioned above, the most effective and practical solution is to set an upper limit for the number of workers that can effectively be employed and moved on the site. The planning and scheduling of the project would then be guided by this upper labour resource limit. It would result in better efficiency of the labour resource, but at the expense of a longer construction period.

4.6.1 Managing the labour resource using histograms and trend graphs

The foregoing discussion has examined problems associated with high peaks of labour demand. In summary, high peaks of demand impact on:

- The quantum of site amenities and safety equipment
- The carrying capacity of personnel hoists, lifts and other equipment needed to transport workers on the site

- Overcrowding of the workplace
- Supervisory needs.

Effective management of the labour resource requires, first, the determination of demand levels over the project period (it is good practice to forecast the labour demand day by day or week by week over the project period), second, the levelling of the resource to achieve its best use, and third, the control of its use.

A linked bar chart serves the purpose of effectively communicating the planning information after it has been generated using the CPM method. One added advantage of a bar chart is that apart from displaying a production schedule, it is also capable of superimposing resource histograms for that particular schedule. This is illustrated in Figure 4.11. The shape of the labour demand histogram assists

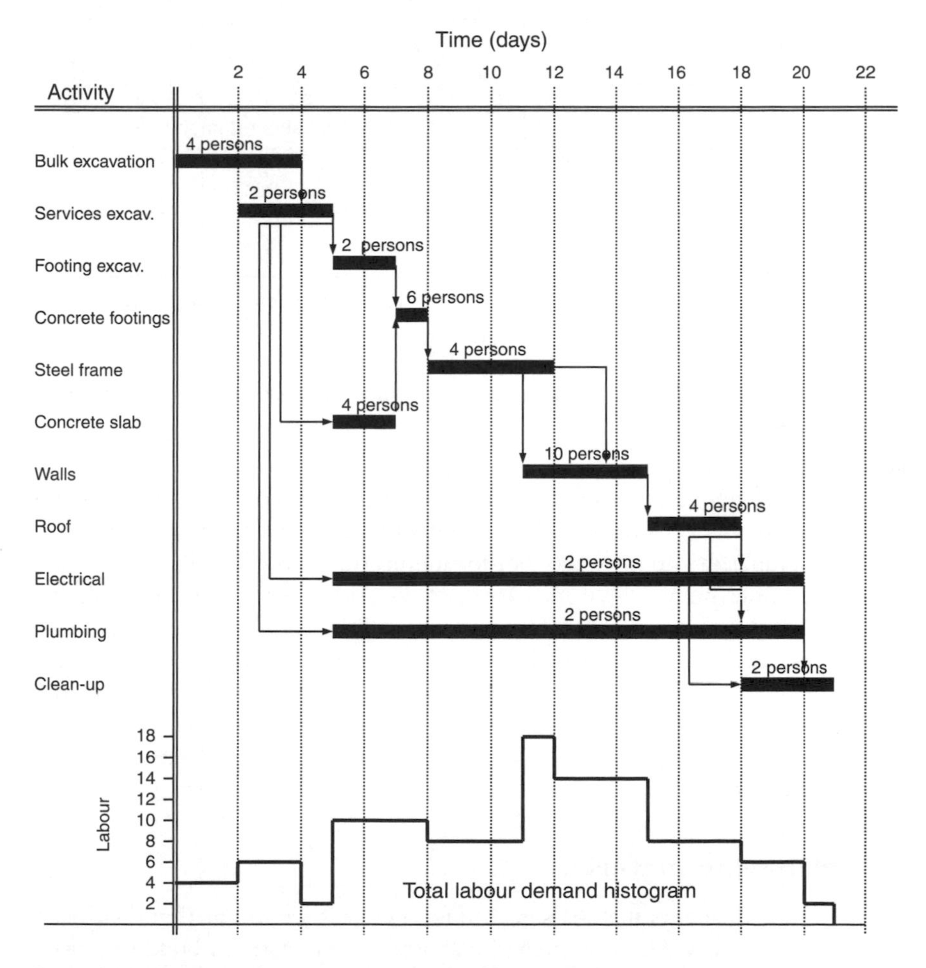

Figure 4.11 The linked bar chart with the total labour demand histogram.

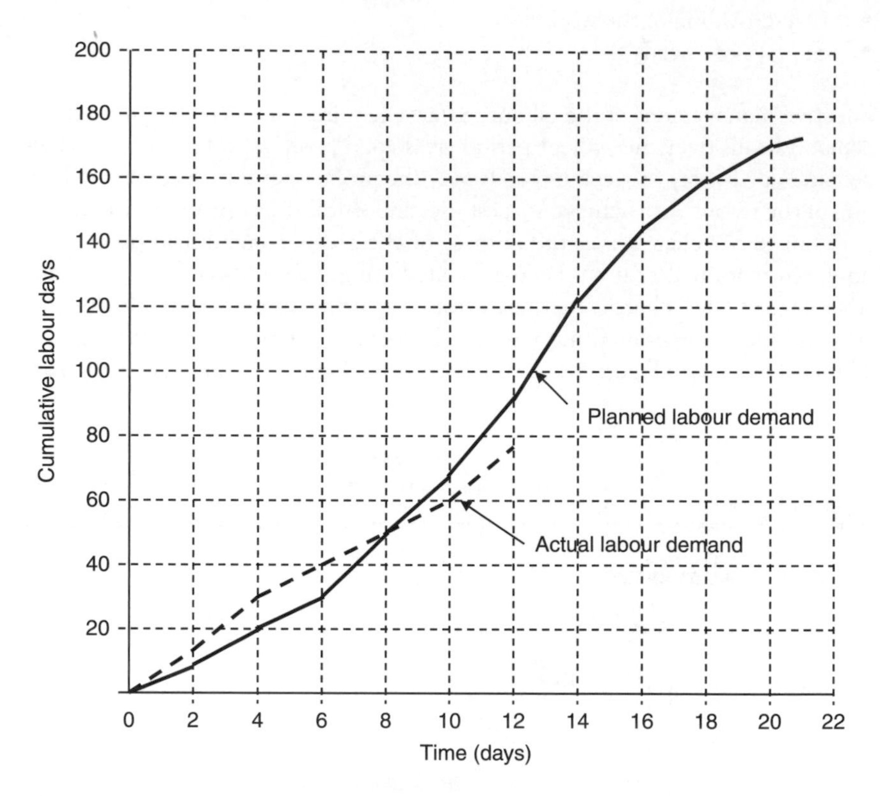

Figure 4.12 Example of a trend graph showing the cumulative planned and actual labour demand.

the planner in resource levelling, in making adequate provisions for amenities and site safety, and in developing efficient system of labour movement on the job.

Trend graphs are an effective tool for monitoring and controlling resources. They are simple, highly visual and easy to create using a spreadsheet. Cumulative total labour demand and cumulative demand trend graphs for different types of labour resources are most common examples of trend graphs. Figure 4.12 illustrates the use of a trend graph in monitoring the cumulative total labour demand. As the name suggests, a trend graph displays the actual demand level of the resource against the planned level. Trends in the level of demand, either positive or negative, can easily be detected and an appropriate action taken.

4.7 Managing materials

The type and quality of materials needed for a project are specified in the design documentation. Irrespective of how the contractor will actually build the structure, the quantity of materials required will be more or less constant. For example, the contractor can build a concrete chimney using either a traditional static formwork method, a slipform method, or a climbing form method. The quantity of reinforcement and concrete will be approximately the same irrespective of the formwork method employed.

Developing a materials schedule for a project is a fairly simple task provided the required materials have been clearly specified, their quantities are known, and supplies are readily available. In Australian urban areas most construction materials are available on short notice, but longer lead-times for their supply in country regions must be allowed for. Lead-times of up to 20 weeks may be required for the delivery of materials purchased overseas and even longer if there is a separate manufacturing process required prior to shipping of the overseas materials.

Two different types of material schedule are commonly prepared:

- A purchasing schedule
- A weekly delivery schedule.

These schedules are prepared in a table format.

A purchasing schedule lists all the required materials, their quantities, the date by which a purchase order must be placed with the supplier, the name of the supplier(s) and the first delivery date. A sample of a purchasing schedule in given in Table 4.2. The important issue here is to ensure that purchase orders have been issued to the suppliers with sufficient lead-times. This is particularly important for imported materials.

Table 4.2 Example of a purchasing schedule of materials

Material	*Unit*	*Total quantity*	*Last date to place purchase order*	*Name of supplier*	*Date of first delivery*	*Order placed*
15MPa Concrete	m^3	500	6.1.10	Pioneer Concrete	10.2.10	Yes
Reinforcement	Tonne	45	15.1.10	Active Steel	1.2.10	Yes
Face bricks	No.	100,000	1.3.10	Austral	3.5.10	Yes
Common bricks	No.	120,000	30.3.10	Austral	26.4.10	Yes
Cement	Tonne	10	12.4.10	Mitre 10 Hardware	26.4.10	Yes
Sand	m^3	30	12.4.10	Mitre 10 Hardware	26.4.10	Yes
Timber (varying sizes)	m	900	26.4.10	Mitre 10 Hardware	18.5.10	Yes
Plasterboard sheets	m^2	800	3.5.10	CSR	24.5.10	Yes
Floor tiles	m^2	150	29.1.10	Italy	18.6.10	Yes
Windows (varying sizes)	No.	120	1.3.10	Stegbar	26.4.10	Yes

Table 4.3 A delivery schedule of materials, week commencing 28 June 2010

Material	*Unit*	*Quantity*	*Name of supplier*	*Date of delivery*	*Delivery confirmed*
15MPa Concrete	m^3	80	Pioneer Concrete	Fri 2.7.10	Yes
Reinforcement	Tonne	7	Active Steel	Mon 28.6.10	Yes
Face bricks	No.	10,000	Austral	Thur 1.7.10	Yes
Common bricks	No.	12,000	Austral	Tue 29.6.10	Yes
Cement	Tonne	2	Mitre 10 Hardware	Mon 28.6.10	Yes
Sand	m^3	5	Mitre 10 Hardware	Mon 28.6.10	Yes
Timber (varying sizes)	m	100	Mitre 10 Hardware	Wed 30.6.10	Yes
Plasterboard sheets	m^2	100	CSR	Thur 1.7.10	Yes
Floor tiles	m^2	35	Italy	Fri 2.7.10	Yes
Windows (varying sizes)	No.	20	Stegbar	Mon 28.6.10	Yes

A weekly delivery schedule shows what required materials are to be delivered, when, in what quantities, and from what suppliers. Such a schedule will be prepared at least a week in advance. It is assumed that the necessary purchase orders have already been placed with the suppliers. Table 4.3 lists the required materials for delivery in the week commencing 28 June 2010.

4.8 Managing plant and equipment

The selection of plant and equipment, particularly the materials-handling plant for use on construction sites, is one of the most important tasks of planners. Like the labour resource, committed plant and equipment are expected to be fully utilised. Some plant, such as compressors, jackhammers, concrete vibrators and the like, is portable. It can easily be moved from activity to activity or returned to the supplier when not needed. This flexibility in mobility helps to ensure the efficient use of plant. Some other plant, such as tower cranes or hoists, may be fixed in place for a specified period, which may even extend over the entire project period. Since such plant is usually cost-intensive, its effective management is essential.

Let's focus on the management of fixed cranes and hoists. The first task that the contractor would need to address is the preparation of the demand histograms for cranes and hoists. These are developed by first identifying those activities that will need to be handled by cranes and hoists. Next, alternative types of suitable cranes and hoists will be selected and their technical specifications and capacities established. Estimates of handling times for each activity for each type of crane and hoist will be made within the schedule of work. Those individual time estimates will then be summed for each day or each week of the schedule. Such information can then be plotted in the form of a demand histogram. Figure 4.13 illustrates a typical demand histogram for cranes and hoists.

The next task is to determine the maximum capacity of each type of crane and hoist. In the example in Figure 4.13, the maximum capacities of a particular crane and a particular hoist are shown as horizontal broken lines. Clearly, neither the crane nor the hoist is able to fully service the scheduled work. With regard to the demand histogram of the crane, the planner may explore a number of possible solutions:

- Selecting a tower crane with a greater capacity that would meet the maximum level of demand
- Using the existing tower crane for the entire project period but boosting its lifting capacity to the maximum demand level between weeks 20 and 35 by committing an additional crane
- Using a mobile crane of the required capacity in the first ten weeks, then replacing it with a tower crane that would meet the maximum demand level between weeks 10 and 40 and then replacing the tower crane with a mobile crane of the required capacity to complete the job between weeks 40 and 50.

If the only crane available to the contractor is the one for which the maximum capacity is shown in Figure 4.13, the contractor would need to reschedule the work within the crane's capacity.

The following example demonstrates a simple approach for determining a lifting demand histogram for a project from which the most appropriate cranes and hoists would be selected. The project in question is a high-rise office building of 40 levels with typical floors between levels 1 and 38. The work has been packaged and let to ten trade contractors. The work of these trade contractors has

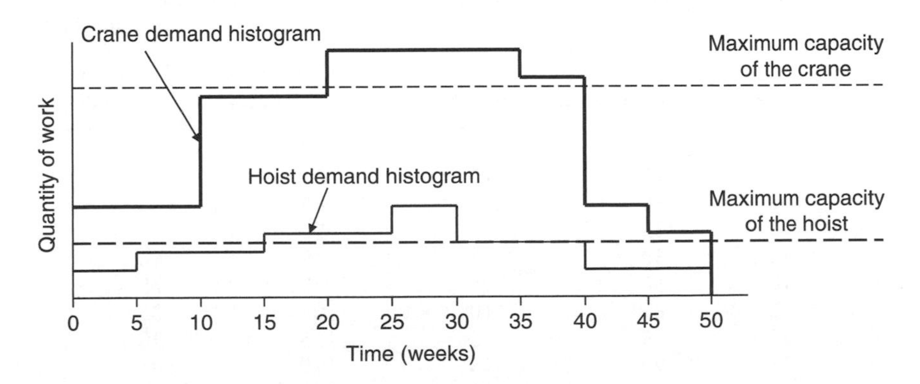

Figure 4.13 Example of a demand histogram for cranes and hoists.

been scheduled sequentially from floor to floor with a cycle time of each work package per floor being one week. It means that only when the first trade contractor reaches level 10 will the tenth trade contractor be able to start work on level 1. From that point onwards, all ten trade contractors would work simultaneously until the work of individual trade contractors comes to an end.

The first task is to identify activities that would need to be handled by cranes or hoists and estimate their handling demand times for different types and capacities of hoisting plant. In this case, the choice of handling plant has been restricted to one specific type of crane and one specific type of hoist (their technical specifications and lifting capacities are known but are not specified here). Since the work has been packaged, handling demand times will be calculated for each package.

This information is given in Table 4.4 in the 'Crane demand time' and the 'Hoist demand time' columns in 'Hours per week'. As the trade contractors work sequentially from floor to floor, the crane and hoist demand times increase from week to week until all ten trade contractors are engaged from week 10 onwards. Thereafter, the lifting demand level will be constant. The progressive build-up of demand for the crane and the hoist is expressed in Table 4.4 in the columns headed 'Cumulative hours/week'.

The cumulative demand histograms of the crane and the hoist are plotted in Figure 4.14. In this example, the maximum capacity of the hoisting plant (for both the crane and the hoist) is set conservatively at 40 hours per week. The cumulative crane demand histogram shows that three cranes would be required to meet the lifting demand of the project. Similarly, the cumulative hoist demand histogram suggests that four hoists would be needed. If the maximum capacity of the crane and the hoist is set at 50 hours per week, two cranes and three hoists would be needed.

Table 4.4 The lifting demand information

	Crane demand time		*Hoist demand time*		*Crane + hoist demand time*
Trades	*Hours per week*	*Cumulative hours/week*	*Hours per week*	*Cumulative hours/week*	*Cumulative hours/week*
1	15	15	20	20	
2	5	20	10	30	
3	10	30	15	45	
4	5	35	5	50	
5	15	50	25	75	60
6	10	60	15	90	75
7	5	65	10	100	85
8	10	75	5	105	90
9	15	90	15	120	105
10	5	95	10	130	115

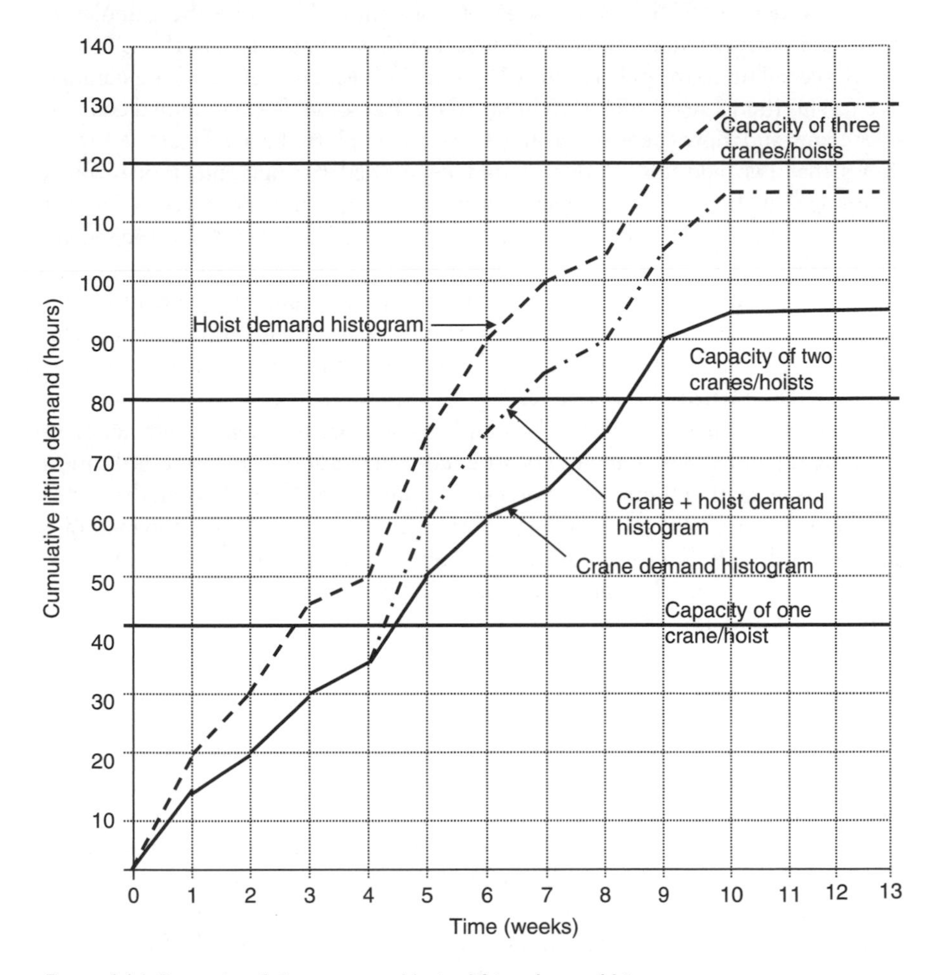

Figure 4.14 Example of the crane and hoist lifting demand histograms.

Most commonly, the planner will select a combination of cranes and hoists. Let's assume that the construction site in question can accommodate one crane only. It means that additional hoists would need to be installed to supplement the crane in order to meet the lifting demand. Let's assume further that the crane is required to service the first three trade contractors and hoists are able to service the other trades.

The cumulative crane demand histogram shows that one crane is able to service up to the first four trade contractors; thereafter, additional hoists would be needed. The number of additional hoists is determined in Table 4.4. The crane must be available for 35 hours per week to service the first four trade contractors (see the 'Cumulative hours/week' column under 'Crane demand time' in Table 4.4). The fifth trade requires 25 hours per week of hoist time. Therefore the cumulative crane and hoist demand time for the first five trade contractors is 35 + 25 = 60 hours/week. The sixth trade requires 15 hours of hoist per week and the cumulative total is therefore 60 + 15 = 75 hours/week, and so on. The combined cumulative crane and hoist demand histogram is then plotted (see Figure 4.14). It shows that two additional hoists would be needed to supplement one crane, assuming that the maximum capacity of the plant is 40 hours per week. The first hoist would need to be installed and made operational on week 4 and the second on week 6.

The final task in effective plant or equipment management is to control efficiency of the plant or equipment that has been committed. Idle times are wasteful and must be avoided at all cost. Plant or equipment that is no longer needed must be promptly returned to the supplier.

A number of useful tools can be employed for monitoring the efficiency of committed plant or equipment. They include calculation of resource utilisation factors, assessment of their utilisation using a multiple activity chart method (this will be examined in detail in Chapter 9), and creation and maintenance of a database of hired plant and equipment. Detailed information on the management of construction plant and equipment can be found in Harris and McCaffer (1991) and Chitkara (1998).

4.9 Summary

This chapter has addressed the issue of resource management. It examined different patterns of resource distribution, particularly those commonly associated with construction projects. It then explained the concept of resource levelling and demonstrated its principles on a simple example. The latter part of the chapter addressed important issues of management of labour, materials and plant/equipment. In the next chapter, the concept of the CPM will be expanded by developing a number of overlapping models for a more realistic scheduling.

Exercises

Solutions to the following exercises can be found on the following website: http://www.routledge.com/books/details/9780415601696/

Exercise 4.1

a) Calculate the precedence schedule given in Figure 4.15. The schedule data are given in Table 4.5.
b) Level the labour resource using the method of trial and error so that the labour demand does not exceed six workers. The project duration cannot be extended.

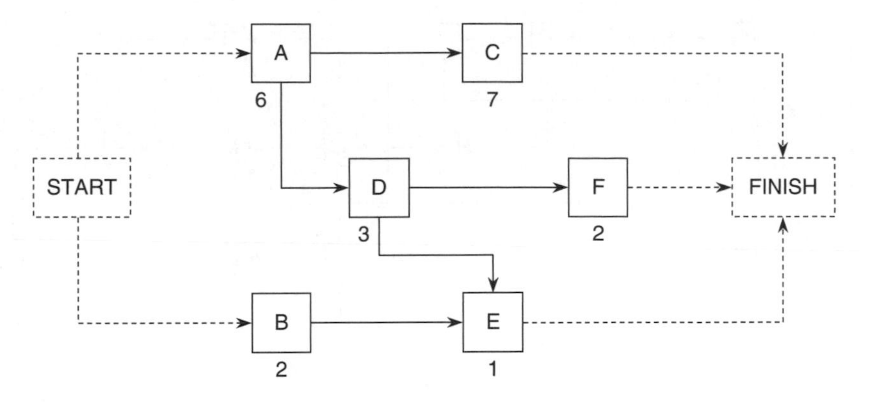

Figure 4.15 The precedence schedule.

Table 4.5 Data for a precedence schedule

Activity	*Time (days)*	*Labour resource rate*	*Labour resource days*
A	6	5	30
B	2	4	8
C	7	2	14
D	3	3	9
E	1	2	2
F	2	2	4
Total			67

Exercise 4.2

a) Calculate the precedence schedule given in Figure 4.16. The schedule data are given in Table 4.6.
b) Level the labour resource using the method of trial and error so that the labour demand does not exceed eight workers. The project duration cannot be extended.

Exercise 4.3

A high-rise commercial building of 34 storeys will be constructed on a one-week cycle per floor. The project will be serviced by tower hoists. Develop a hoist demand schedule from the information in Table 4.7. How many hoists will be required? Add 10 per cent for contingencies.

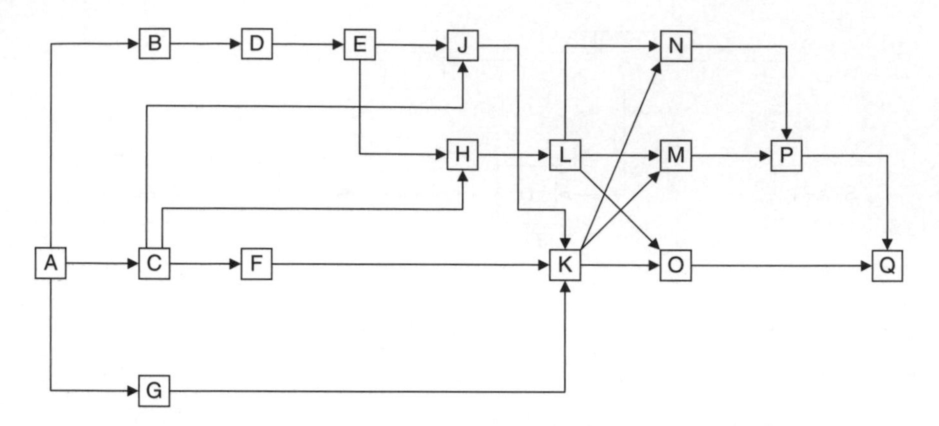

Figure 4.16 The precedence schedule.

Table 4.6 Data for a precedence schedule

Activity	*Time (days)*	*Labour resource rate*	*Labour resource days*
A	2	4	8
B	2	5	10
C	8	1	8
D	3	3	9
E	1	1	1
F	10	3	30
G	8	3	24
H	3	5	15
J	3	2	6
K	4	4	16
L	4	2	8
M	1	3	3
N	2	4	8
O	8	2	16
P	2	2	4
Q	2	3	6
Total			*172*

Exercise 4.4

A 34-storey high-rise commercial building will require one tower crane for materials-handling. The requirement of one tower crane has been predetermined using a crane demand schedule and by reviewing the site layout. The tower crane will be positioned within the only goods lift shaft and removed once the structure is complete. The crane will be used to supply materials to construct a concrete

Table 4.7 Activities to be performed by a crane

Trade contract	*Activity*	*No. of loads/ floor*	*Cycle/ floor (minutes)*	*Activity time/ floor (hours)*	*Total time/ floor (hours)*	*Cumulative time (hours)*
1	Formwork	100	15			
	Contingency	10%				
2	Reinforcement	40	15			
	Concrete	170	7			
	Conduits and cables	5	30			
	Contingency	10%				
3	Handrails	6	15			
	Contingency	10%				
4	A/C ducts	20	15			
	Sprinkler pipes	10	15			
	Contingency	10%				
5	Plumbing stock	5	30			
	Lift rails	3	30			
	Contingency	10%				
6	Bricks	15	15			
	Mortar	10	15			
	Windows	7	60			
	Door frames	3	30			
	Contingency	10%				
7	Electrical	8	60			
	Plaster	30	15			
	Glazing	8	60			
	Contingency	10%				
8	Ceiling frames	4	30			
	Wall and floor tiles	20	20			
	Contingency	10%				
9	Toilet partitions	2	30			
	Contingency	10%				
10	Plumbing fixtures	2	60			
	Contingency	10%				
11	Ceiling tiles	8	30			
	Lights	6	60			
	Contingency	10%				
12	Lift doors	17	30			
	Contingency	10%				
13	Doors	2	30			
	Vanity units	3	60			
	Venetian blinds	1	60			
	Mirrors	3	60			
	Contingency	10%				
14	Induction units	2	30			
	Lift lobby finish	12	20			
	Door hardware	4	15			
	Contingency	10%				

structure comprising columns, beams, slabs, walls, lift shafts and one goods lift shaft. The crane supplier has given the option of three cranes and specifications for each. Select the preferred crane for the project. Costs analysis is not required for this exercise.

Project information

PROJECT INFORMATION REQUIRED FOR CRANE SELECTION:

1 The lift shaft will be built to level 3 (three storeys) prior to installation of the crane and will take four weeks to complete.
2 The crane will be installed within the only goods lift.
3 The jump-form system will be installed using the crane and will take three weeks. The jump-form is a mechanical self-climbing formwork system for use in construction of the lift shafts.
4 Thirty-four floors of the structure will take 35 weeks to complete once the jump form is installed. There are approximately 329 load lifts required per floor with the average load weighing 4 t.
5 The jump-form system removal can take place after the structure is complete and will take three weeks.
6 The roof plant room (level 35) is to be constructed from structural steel with the largest steel member weighing 5 t. It is located 45 m from the goods lift shaft.
7 The heaviest permanent plant to be lifted to the roof plant room weighs 7 t and is located 40 m from the goods lift shaft.
8 The crane can be removed after the final piece of plant is lifted into position.

Assume: Eight-hour work days, six-day work weeks.
Note: Once tested, the final crane loading capacity is to be confirmed by the crane supplier and structural engineer.

CRANE 1

- Crane erection – 5 days
- Maximum capacity – 16 t
- 8.25 t capacity at a maximum reach of 60 m radius
- Crane dismantle/removal – 6 days
- Average cycle time per lift – 12 minutes.

CRANE 2

- Crane erection – 3 days
- Maximum capacity – 12 t
- 6 t capacity at a maximum reach of 50 m radius

- Crane dismantle/removal – 4 days
- Average cycle time per lift – 9 minutes.

CRANE 3

- Crane erection – 3 days
- Maximum capacity – 8 t
- 5 t capacity at a maximum reach of 50 m radius
- Crane dismantle/removal – 4 days
- Average cycle time per lift – 8 minutes.

Chapter 5

Overlapping network models

5.1 Introduction

This chapter aims to expand a sequential, finish-to-start (FTS) relationship between preceding and succeeding activities in a critical path schedule by developing a broad spectrum of possible overlapped relationships.

The discussion of the critical path method (CPM) has thus far assumed that for a pair of activities of which one is preceding and the other succeeding, the preceding activity must be fully completed before the succeeding activity can proceed. This relationship is referred to as FTS. The assumption that the preceding activity must be fully completed before the succeeding one could begin was made in Chapter 3 to simplify the task of defining and developing the concept of the CPM.

Now that the concept of the CPM has been defined and developed, it is necessary to remove the assumption that work can be scheduled only sequentially. Many activities in a schedule can be overlapped in order to speed up the production process. It is advantageous for the planner to start the succeeding activity, where possible, before its preceding activity has been fully completed. For example, placing of slab reinforcement usually starts before the deck formwork has been fully completed. Overlapping not only models relationships between activities more realistically but it also reduces the overall duration of a schedule and allows the early deployment of resources.

In the following sections of this chapter, the concept of the CPM will be expanded by defining a number of overlapping models. These are:

- A start-to-start (STS) link
- A finish-to-finish (FTF) link
- A start-to-finish (STF) link
- Combination of STS and FTF commonly referred to as a compound link.

A pair of activities, one preceding and the other succeeding, can thus have one of five possible links: the traditional FTS link and the four overlapping links. Each of these links will now be closely examined.

5.2 Finish-to-start link (FTS)

The FTS link is most frequently used in critical path schedules. It is the link around which the concept of the CPM was defined in Chapter 3. In any pair of preceding and succeeding activities, the succeeding activity, J, may either start immediately after the completion of the preceding activity, I, or some time later. The start of the succeeding activity, J, may be delayed by unavailability of resources. Its start may also be delayed to allow the activity to gain the required physical properties through curing. For example, if the activity 'Render walls' in Figure 5.1(a) is completed today, the succeeding activity 'Paint walls' cannot start until the render has sufficiently cured, which is commonly specified as 14 days. The planner would need to treat this delay caused by curing as an extra activity in the schedule (see Figure 5.1(b)). Rather than treating delays caused by events like curing as extra activities, the planner may simply express delay as lead-time and note it on the relevant link in the schedule (see Figure 5.1(c)). When 'lead-time' replaces a 'delay activity', the formula for calculating the earliest start date (ESD) of the succeeding activity, J, for the n number of preceding I activities is modified as follows:

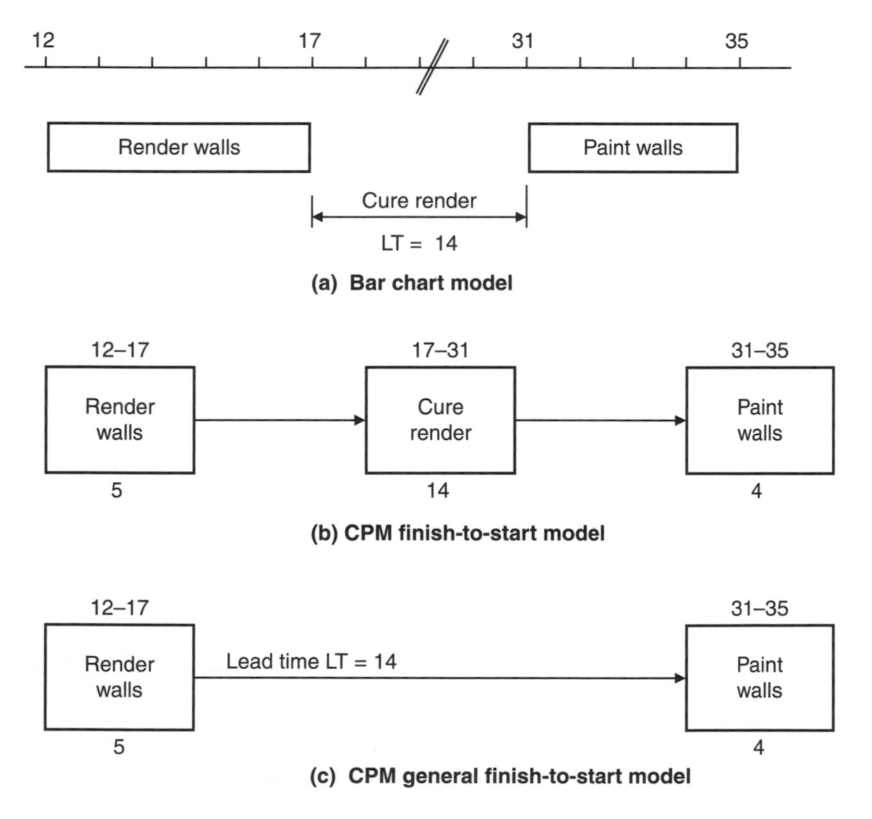

Figure 5.1 Example of the finish-to-start link.

$$ESD_J = MAX\ (EFD_I + LT_I)$$

Similarly, the expression of lag for the link IJ with lead-time becomes:

$$Lag_{IJ} = ESD_J - EFD_I - LT_I$$

5.3 Start-to-start link (STS)

For a pair of activities, I and J, the succeeding activity, J, may sometimes begin after a portion of the preceding activity, I, has been completed. Beyond this point, activities I and J are totally independent of each other. Such a link between activities I and J is referred to as the STS link.

Let's illustrate this overlap link on a simple example in Figure 5.2(a). The succeeding activity, 'Footings to service core', is scheduled to start three

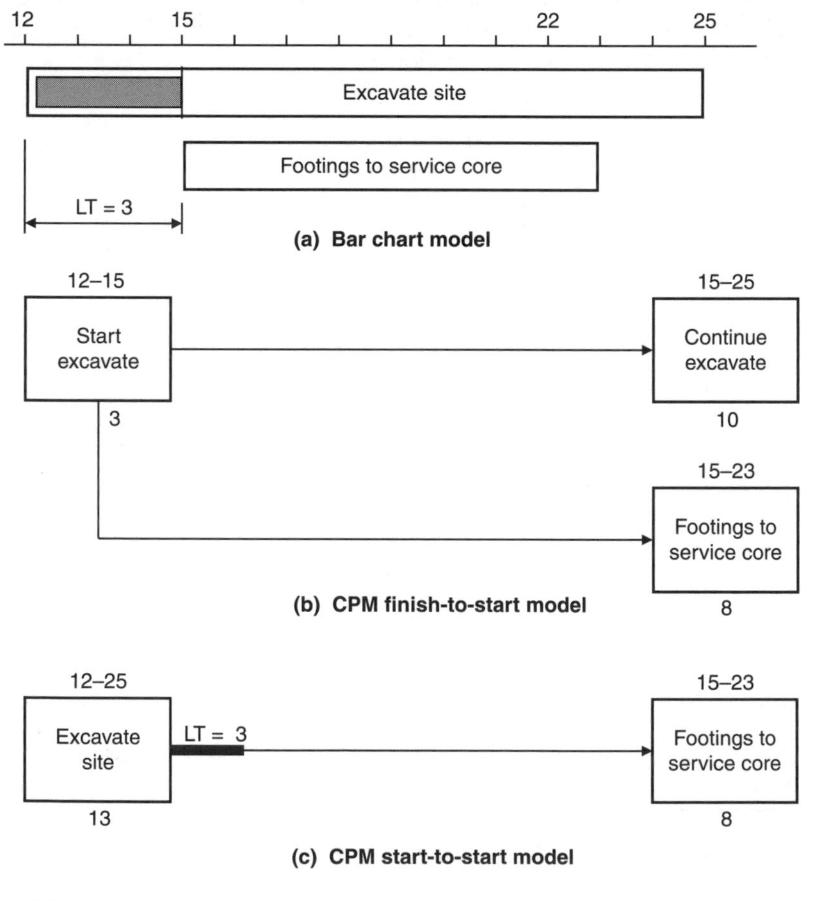

Figure 5.2 Example of the start-to-start link.

time-units after the preceding activity, 'Excavate site', has commenced. Thereafter, the work associated with these two activities will progress independently of each other and the finish of the preceding activity, 'Excavate site', will have no effect on the finish of the succeeding activity, 'Footings to service core'. This point is reinforced in Figure 5.2(a) where the preceding activity, 'Excavate site', is scheduled to finish after the succeeding activity, 'Footings to service core', has been completed.

In the STS link, overlap occurs in the preceding activity, I ('Excavate site'), and is expressed as lead-time $LT_I = 3$. The bar chart in Figure 5.2(a) clearly shows the nature of the STS link.

If only FTS links are used in a precedence schedule, the relationship between the activities 'Excavate site' and 'Footings to service core' would need to be expressed with three activities as shown in Figure 5.2(b), where the activity 'Excavate site' is split into two separate activities, 'Start excavate' and 'Continue excavate'.

However, using the STS link, only two activities are needed in a precedence schedule to show the relationship between the activities 'Excavate site' and 'Footings to service core' (see Figure 5.2(c)). For clarity, a simple protocol is adopted for identifying STS links. It requires a portion of the link near activity I, where the overlap has occurred, to be widened. The amount of overlap is noted above the widened portion of the link as lead-time LT (Harris 1978).

In the STS link, the ESD of the succeeding activity, J, is calculated first from the ESD of the preceding activity, I, plus LT_I. Formulas for calculating the ESD of the following activity, J, and lag between the activities IJ are expressed as:

$$ESD_J = MAX\ (ESD_I + LT_I)$$

$$Lag_{IJ} = ESD_J - ESD_I - LT_I$$

The validity of these formulas can easily be verified on the bar chart in Figure 5.2(a).

The STS link occurs on rare occasions only and should not be confused with the compound link that is characterised by the presence of both the STS and the FTF (finish-to-finish) links between a pair of preceding and succeeding activities. The compound link will be discussed later in this chapter.

5.4 Finish-to-finish link (FTF)

When two activities have a FTF link, the succeeding activity, J, cannot be completed until its preceding activity, I, has been fully accomplished. This relationship is shown in a bar chart format in Figure 5.3(a). It is important to note that both activities I and J can start totally independently of each other. This is emphasised in Figure 5.3(a) by having the succeeding activity, 'Install mechanical ventilation and test', starting ahead of its preceding activity, 'Power to mechanical ventilation'. In the FTF link, overlap occurs in the succeeding activity, J.

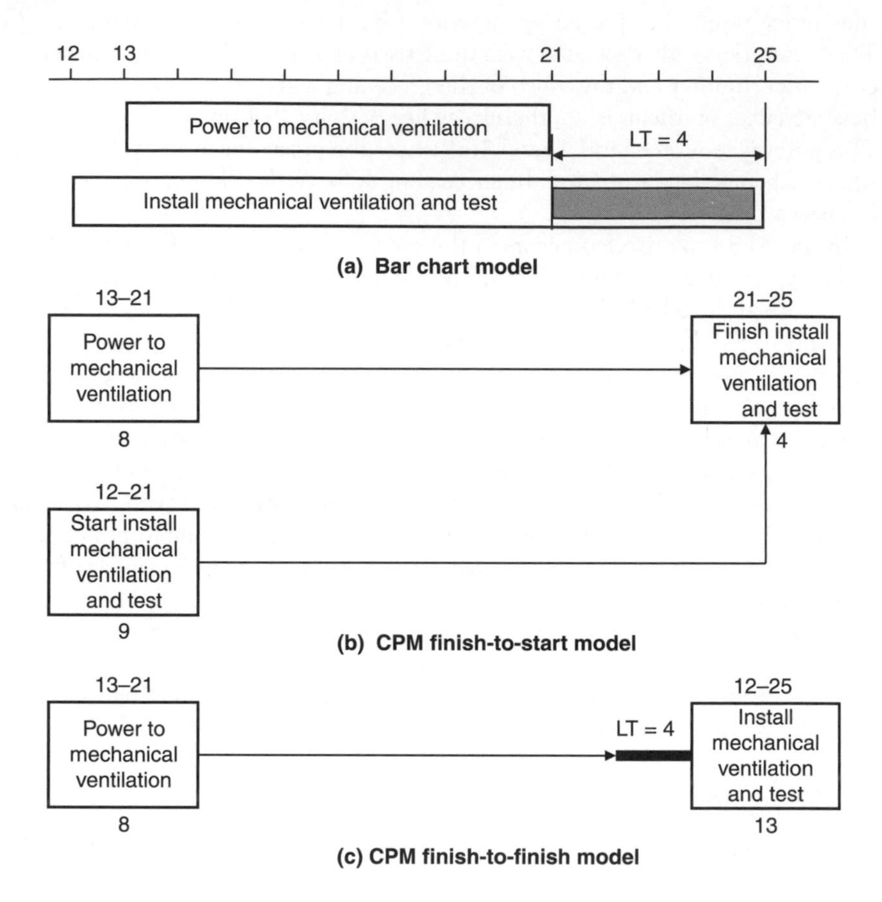

Figure 5.3 Example of the finish-to-finish link.

If the relationship between the activities 'Power to mechanical ventilation' and 'Install mechanical ventilation and test' is shown in a precedence schedule using the FTS link, the succeeding activity 'Install mechanical ventilation and test' would need to be split into two separate activities, 'Start install mechanical ventilation and test' and 'Finish install mechanical ventilation and test'. In total, a schedule would have three activities; see Figure 5.3(b).

Using the FTF link, a precedence schedule would have only two activities, as shown in Figure 5.3(c). The widened portion of the link near activity J signifies the presence of the FTF link. The amount of overlap is noted as a lead-time.

The FTF link requires that the earliest finish date (EFD) of the succeeding activity, J, be calculated first from the EFD of the preceding activity, I, with the addition of LT_J. The ESD of activity J is then calculated as EFD_J – $duration_J$.

Formulas for calculating the EFD of the following activity, J, and lag between activities IJ are expressed as:

$$EFD_J = MAX\ (EFD_I + LT_J)$$

$$Lag_{IJ} = EFD_J - EFD_I - LT_J$$

The validity of the above formulas can easily be verified on the bar chart in Figure 5.3(a).

5.5 Start-to-finish link (STF)

The STF link represents a complex relationship between a pair of preceding and succeeding activities, I and J. The main characteristic of the STF link is that activities I and J start and finish independently of each other. However, they come together at a certain point in their duration where a lead-time portion of the preceding activity I must be completed so that a lead-time portion of the succeeding activity J can be accomplished. This relationship is shown graphically in the bar chart in Figure 5.4(a).

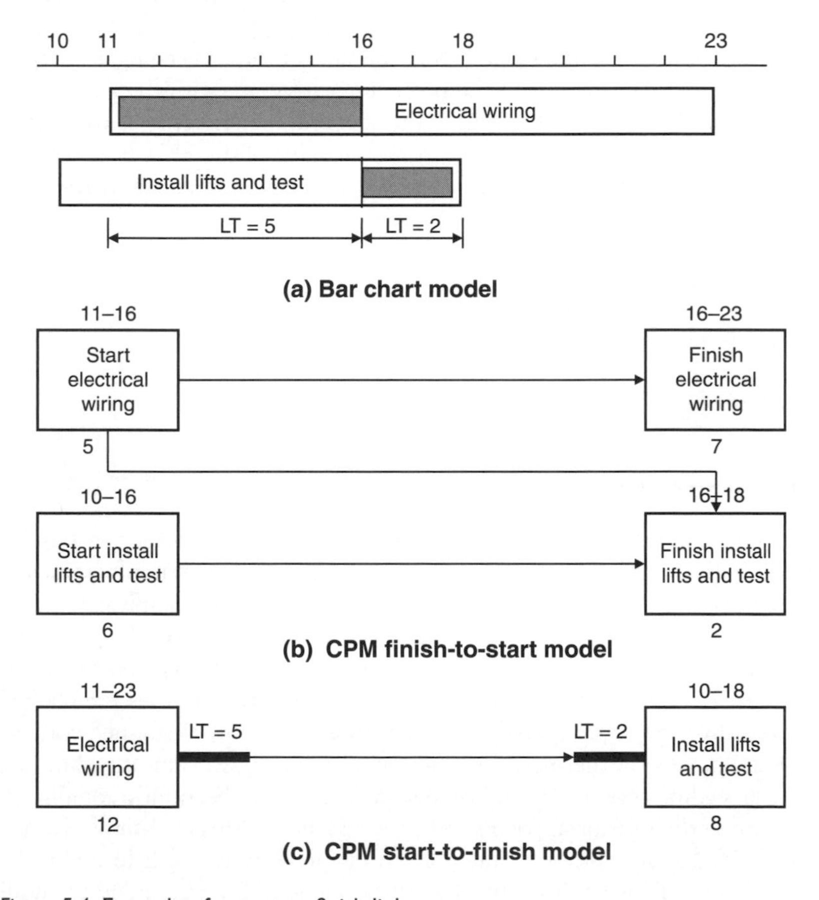

Figure 5.4 Example of a start-to-finish link.

The STF link frequently occurs between activities referred to as 'services'. Such activities typically include 'Mechanical', 'Electrical', 'Hydraulic', 'Fire' and 'Vertical transport' services. For example, the activity 'Electrical wiring' often starts near the beginning of the project and continues almost up till the project's completion. Other 'services' activities such as 'Installation of lifts', 'Air conditioning', 'Hydraulics', 'Fire protection' and the like proceed independently from the activity 'Electrical wiring', but at a specific point in their duration they require access to electric power so that their installation and testing can be completed.

The relationship between the activities 'Electrical wiring' and 'Install lifts and test' could be modelled using FTS links. However, each activity must first be split into two in order to correctly represent the link between those two activities. This is illustrated in Figure 5.4(b).

Using the STF link, a precedence schedule only requires two activities to define the relationship between the activities 'Electrical wiring' and 'Install lifts and test' (see Figure 5.4(c)). The STF link generates two overlaps, one related to the activity 'Electrical wiring' and the other to the activity 'Install lifts and test'. They are noted as lead-times.

The nature of the STF link requires that the EFD of the succeeding activity, J, is calculated first from the ESD of the preceding activity, I, with the addition of both lead-times $LT_I + LT_J$. The ESD of the succeeding activity, J, is calculated next by deducting duration of activity J from the value of the EFD. Formulas for calculating the EFD of the following activity, J, and lag between activities IJ are expressed as:

$$EFD_J = MAX\ (ESD_I + LT_I + LT_J)$$

$$Lag_{IJ} = EFD_J - ESD_I - LT_I - LT_J$$

The bar chart in Figure 5.4(a) confirms the validity of the formulas.

5.6 A compound link

A compound link between a pair of preceding and succeeding activities, I and J, consists of both the STS and the FTF links. This is the most commonly occurring overlap and is illustrated in the bar chart in Figure 5.5(a). This graph shows that the succeeding activity, 'Wall lining', cannot start until a portion of the preceding activity, 'Install timber stud walls', has been completed, which is given by a lead-time of four. This part of the relationship between the two activities is modelled by the STS link. From this point onwards, both activities are performed concurrently, but a portion of the succeeding activity, 'Wall lining', cannot be finished until the preceding activity, 'Install timber stud walls', has been fully completed. This portion of the relationship is given by the FTF link with a lead-time of three.

Figure 5.5(a) shows the lead-time portion of the activity 'Wall lining' to be three time-units. It means that the last three time-units of the activity 'Wall

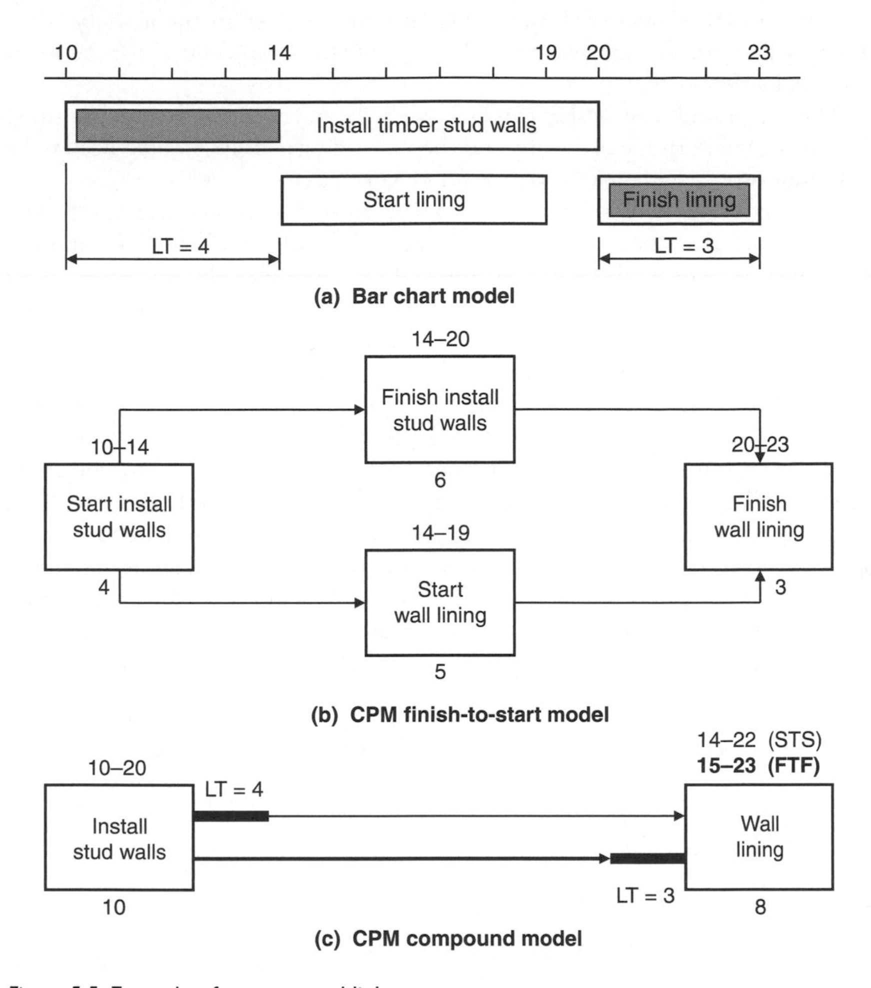

Figure 5.5 Example of a compound link.

lining' cannot start until the preceding activity, 'Install timber stud walls', has been completed. It should now be clear that the activity 'Wall lining' will be delayed by one time-unit. To correctly schedule the activity 'Wall lining', the planner would need to either split it into two, as shown in Figure 5.5(a), with a one-time-unit break between the two parts, or delay its start by one time-unit to maintain continuity of work. Let's examine the former alternative first.

The contractor may want to begin the succeeding activity, 'Wall lining', as soon as the lead-time of the preceding activity, 'Install timber stud walls', has been reached. The work of the two activities then proceeds in parallel. However, if the rate of progress of the subcontractor performing the succeeding activity, 'Wall lining', is greater than that of the subcontractor performing the preceding activity, 'Install timber stud walls', discontinuity of work in the activity 'Wall lining' will occur. For the example in Figure 5.5(a), discontinuity occurs between time-units 19 and 20.

The compound relationship can be modelled in a critical path schedule using FTS links, but both the preceding and the succeeding activities would need to be split into two activities each, as shown in Figure 5.5(b).

Alternatively, the compound link can be created in a precedence schedule with two activities only, as shown in Figure 5.5(c) with both the STS and the FTF links present. Calculations of the ESD of the succeeding activity, 'Wall lining', will be performed separately for each of two links using the formulas derived in sections 5.3 and 5.4. For the link FTF, the value of the EFD of the activity 'Wall lining' will be calculated first, and the value of the ESD will be derived from this. The greater of the two values of the ESD is the ESD of the activity 'Wall lining'. In the example in Figure 5.5(c), the FTF link is dominant. However, if the lead-time of the activity 'Wall lining' is reduced to one time-unit, the STS link would then be dominant. If the lead-time of the activity 'Wall lining' is two time-units, both links have the same degree of dominance.

It was noted earlier that continuity of work in the activity 'Wall lining' would be ensured by delaying its start by one time-unit. It is interesting to observe that when a compound link is calculated (see Figure 5.5(c)), the succeeding activity, 'Wall lining', is automatically scheduled in a manner that ensures its continuity.

5.7 Free and total float in overlapped networks

The formulas for free and total float (TF) derived in Chapter 3 are valid for use in overlapped networks. These formulas are:

$$\text{Free float}_I = \text{MIN}\ (\text{Lag}_{I\,J_1, J_2 \ldots J_n})$$

$$\text{Total float}_I = \text{MIN}\ (\text{Lag}_{I\,J_1, J_2 \ldots J_n} + \text{TF}_J)$$

In overlapped networks it is important to distinguish between the end activity and the terminal activity. The 'end' activity is the last activity in a CPM schedule, which may or may not determine its overall duration. On the other hand, the terminal activity may or may not be the last activity in a CPM schedule but it determines its overall duration. In networks that use FTS links only, end activities are those that have no further succeeding links. One or possibly more end activities will be critical and will thus determine the overall duration of the network. The end activity or activities that are critical are also 'terminal'.

However, in overlapped networks, the end activity may have no further succeeding links, yet it may not be critical. In such a case, the end activity is not terminal since it does not determine the overall duration of the project; another activity will determine the overall network duration, which will be a true terminal activity. In Figure 5.6, activity H is the last activity in the schedule, but when the

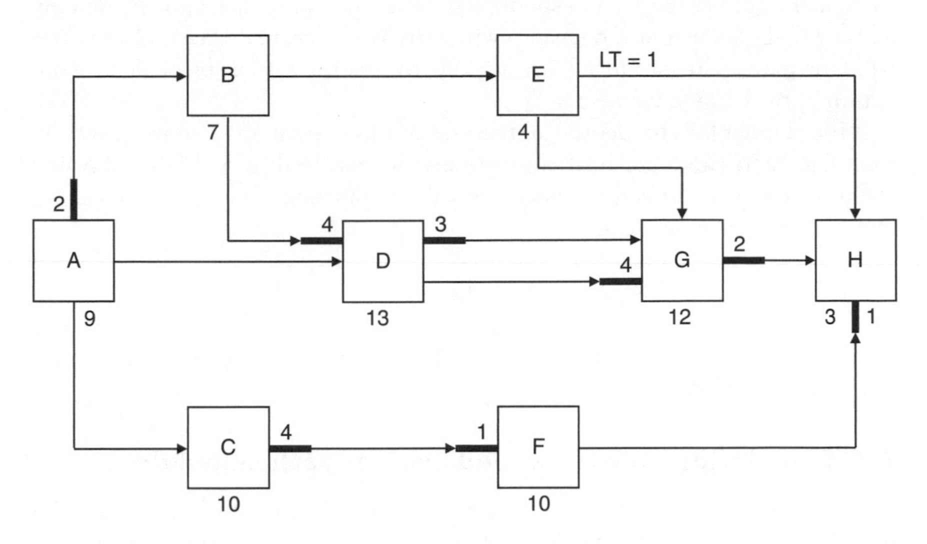

Figure 5.6 Example of an overlapped precedence schedule.

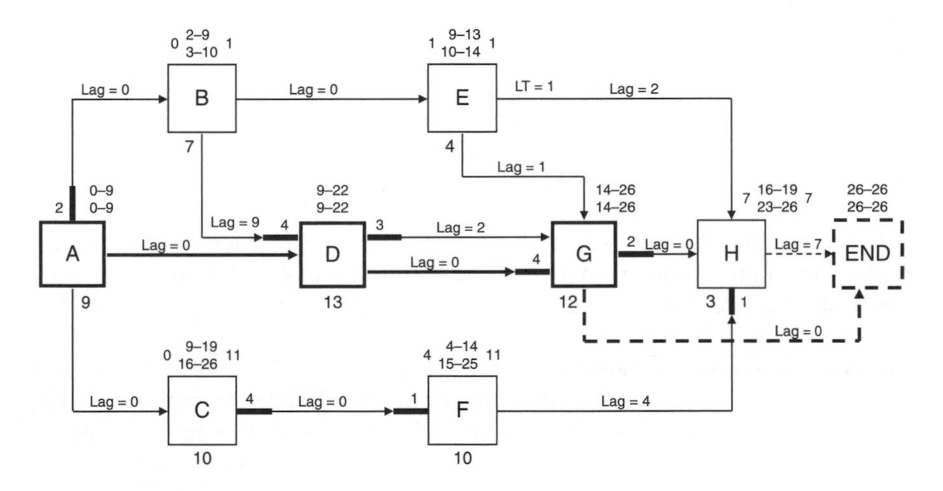

Figure 5.7 The fully calculated overlapped precedence schedule.

schedule is calculated, the true terminal activity that determines the schedule's duration is activity G. This is illustrated in Figure 5.7. The reason why activity G is terminal will be discussed in section 5.8.

Overlapped precedence networks are often difficult to interpret. For example, it is not clearly apparent in Figure 5.7 that activity C is one of the end activities in the schedule. On closer examination of the schedule, it becomes obvious that activity C is in fact one of three end activities and as such it should be linked to the dummy activity END, and should have the value of its lag with the dummy activity END calculated. The reason why activity C is one of three end activities will be explained in section 5.8. Meanwhile, the reader may be able to deduce the reason by referring to Figure 5.9.

If the planner fails to identify all the relevant links in an overlapped schedule, the values of free and total float of some activities may be distorted. The following formula should be applied to check the validity of calculated float values in an overlapped precedence schedule:

$$FF_I \text{ and } TF_I \leq (LFD_{\text{Terminal activity}} - EFD_I)$$

The following example will demonstrate the calculation process applied to an overlapped schedule.

5.8 Calculating an overlapped critical path schedule

The calculation process applied to overlapped critical path schedules follows the procedure defined in Chapter 3 under the concept of link lag (section 3.6). It involves the following steps:

- Forward pass calculations
- Calculation of lags
- Identification of a critical path
- Calculation of free and total float
- Calculation of latest start dates (LSDs) and latest finish dates (LFDs)
- Plotting of a linked bar chart.

This calculation process will now be applied to a precedence schedule in Figure 5.6. Links between individual activities of the schedule represent the entire array of possible overlaps.

Step 1: Forward pass calculations

The ESDs and EFDs of the activities in the schedule are calculated using formulas derived for each link type. Individual calculations are given below:

Link AB (STS):

$ESD_B = ESD_A + LT_A = 0 + 2 = 2$ $\qquad$ $EFD_B = 2 + 7 = 9$

Link AC (FTS):

$ESD_C = EFD_A = 9$ $\qquad$ $EFD_C = 9 + 10 = 19$

Activity D has two preceding links, one of which will determine its ESD and EFD values:

Link AD (FTS):

$ESD_D = EFD_A = 9$ $EFD_B = 9 + 13 = 22$

Link BD (FTF):

$EFD_D = EFD_B + LT_D = 9 + 4 = 13$ $ESD_D = 13 - 13 = 0$

Therefore, activity D has ESD = 9 and EFD = 22

Link BE (FTS):

$ESD_E = EFD_B = 9$ $EFD_E = 9 + 4 = 13$

Link CF (STF):

$EFD_F = ESD_C + LT_C + LT_F = 9 + 4 + 1 = 14$ $ESD_F = 14 - 10 = 4$

Activity G has three preceding links, one of which will determine its ESD and EFD values:

Link EG (FTS):

$ESD_G = EFD_E = 13$ $EFD_G = 13 + 12 = 25$

Link DG (STS):

$ESD_G = ESD_D + LT_D = 9 + 3 = 12$ $EFD_G = 12 + 12 = 24$

Link DG (FTF):

$EFD_G = EFD_D + LT_G = 22 + 4 = 26$ $ESD_G = 26 - 12 = 14$

Therefore, activity G has ESD = 14 and EFD = 26.

Activity H also has three preceding links, one of which will determine its ESD and EFD values:

Link EH (FTS):

$ESD_H = EFD_E + LT_E = 13 + 1 = 14$ $EFD_H = 14 + 3 = 17$

Link GH (STS):

$ESD_H = ESD_G + LT_G = 14 + 2 = 16$ $EFD_H = 16 + 3 = 19$

Link FH (FTF):

$EFD_H = EFD_F + LT_F = 14 + 1 = 15$ $\qquad ESD_H = 15 - 3 = 12$

Therefore, activity H has ESD = 16 and EFD = 19.

The fully calculated schedule is given in Figure 5.7.

Although activity H is the last activity in the schedule, activity G is the true terminal activity. Since activities H and G are both end activities, they have links to a dummy activity, END.

Step 2: Calculate lags

Formulas derived for each link type are used to calculate lag values, which are as follows:

Link AB (STS) $ESD_B - ESD_A - LT_A = 2 - 0 - 2 = 0$

Link AD (FTS) $ESD_D - EFD_A = 9 - 9 = 0$

Link AC (FTS) $ESD_C - EFD_A = 9 - 9 = 0$

Link BE (FTS) $ESD_E - EFD_B = 9 - 9 = 0$

Link BD (FTF) $EFD_D - EFD_B - LT_D = 22 - 9 - 4 = 9$

Link CF (STF) $EFD_F - ESD_C - LT_C - LT_F = 14 - 9 - 4 - 1 = 0$

Link DG (STS) $ESD_G - ESD_D - LT_D = 14 - 9 - 3 = 2$

Link DG (FTF) $EFD_G - EFD_D - LT_G = 26 - 22 - 4 = 0$

Link EH (FTS) $ESD_H - EFD_E - LT_E = 16 - 13 - 1 = 2$

Link EG (FTS) $ESD_G - EFD_E = 14 - 13 = 1$

Link FH (FTF) $EFD_H - EFD_F - LT_F = 19 - 14 - 1 = 4$

Link GH (STS) $ESD_H - ESD_G - LT_G = 16 - 14 - 2 = 0$

Link G–END (FTS) $ESD_{END} - EFD_G = 26 - 26 = 0$

Link H–END (FTS) $ESD_{END} - EFD_H = 26 - 19 = 7$

Step 3: Determine the critical path

The critical path in a precedence schedule is a path of zero lags. The schedule in Figure 5.7 has only one critical path connecting activities A, D and G. The critical link between activities D and G is the FTF link.

Step 4: Calculate free and total float

Free and total float of the non-critical activities are calculated using the formulas derived in section 5.7 above. Free floats have already been calculated in the form of lags. They are as follows:

$FF_B = 0$ (the minimum lag value)

$FF_C = 0$

$FF_E = 1$ (the minimum lag value)

$FF_F = 4$

$FF_H = 7$

Total floats of non-critical activities are calculated by working from the terminal activity and other critical activities back to the beginning of the schedule. Total float calculations are given below.

$TF_H = TF_{END} + Lag_{H\text{–}END} = 0 + 7 = 7$

$TF_F = TF_H + Lag_{FH} = 7 + 4 = 11$

$TF_E = TF_H + Lag_{EH} = 7 + 2 = 9$, or

$TF_G + Lag_{EG} = 0 + 1 = 1$, therefore, $TF_E = 1$ (it is the minimum value)

$TF_C = TF_F + Lag_{CF} = 11 + 0 = 11$

$TF_B = TF_E + Lag_{BE} = 1 + 0 = 1$, or

$TF_D + Lag_{BD} = 0 + 9 = 9$, therefore, $TF_B = 1$ (it is the minimum value)

The next step is to check the validity of the values of free and total float using the formula given in section 5.7 as:

FF_I and $TF_I \leq (LFD_{TERM.} - EFD_I)$.

Then,

$FF_B = 26 - 9 = 17$ (OK), $TF_B = 26 - 9 = 17$ (OK)

$FF_C = 26 - 19 = 7$ (OK), $TF_C = 26 - 19 = 7$ (This is the maximum value of TF)

$FF_E = 26 - 13 = 13$ (OK), $TF_E = 26 - 13 = 13$ (OK)

$FF_F = 26 - 14 = 12$ (OK), $TF_F = 26 - 14 = 12$ (OK)

$FF_H = 26 - 19 = 7$ (OK), $TF_H = 26 - 19 = 7$ (OK, right on the limit)

The total float value of activity C must be reduced to 7. The reason why the originally calculated value of total float is incorrect is because activity C is in fact an end activity and its link with the dummy activity END is missing. Careful examination of the STF link between activities C and F reveals the reason why. Because the end of activity C is not linked to any other activity, it must be an end activity. This can be confirmed by examining the bar chart in Figure 5.9.

The missing link C–END has been added to the schedule and its lag value calculated as 7 (see Figure 5.8). With this additional link, the value of total float of activity C is then 7 (for the link C–END).

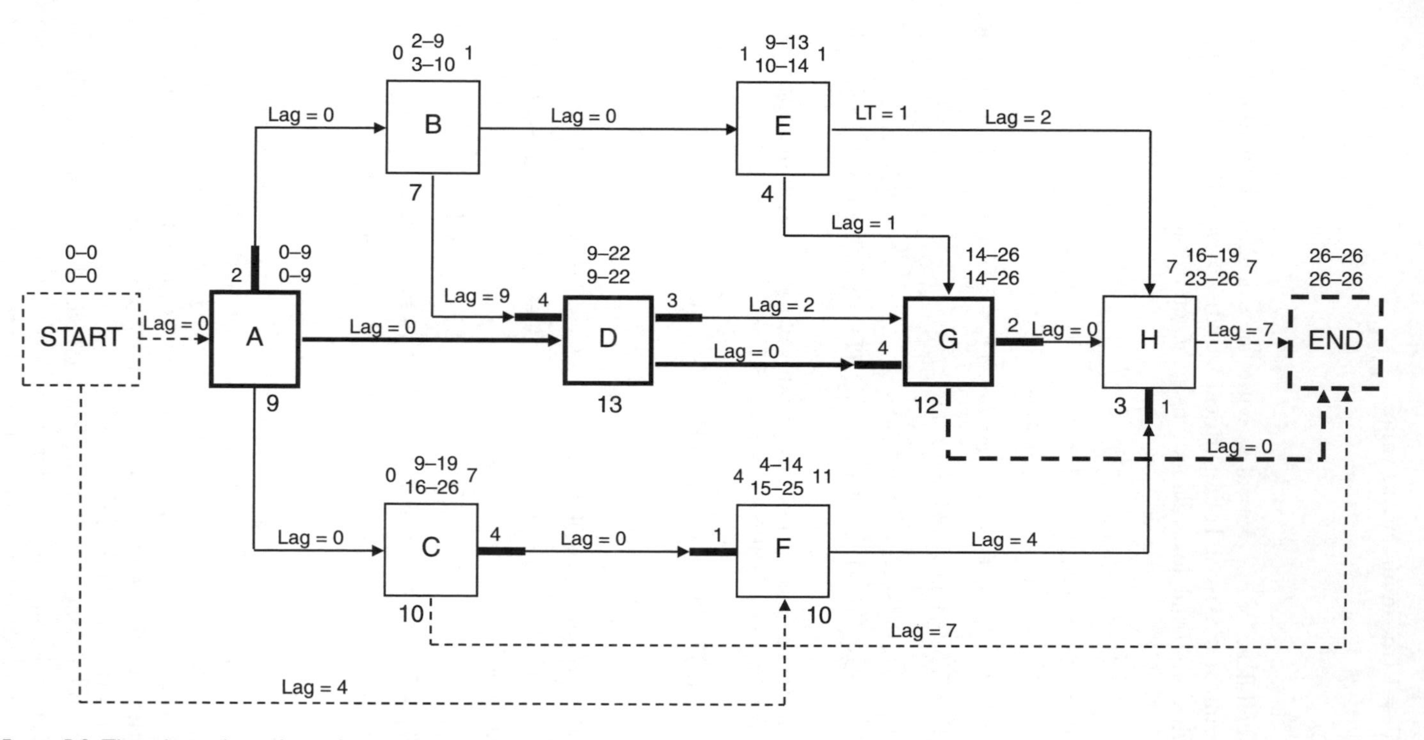

Figure 5.8 The adjusted overlapped precedence schedule.

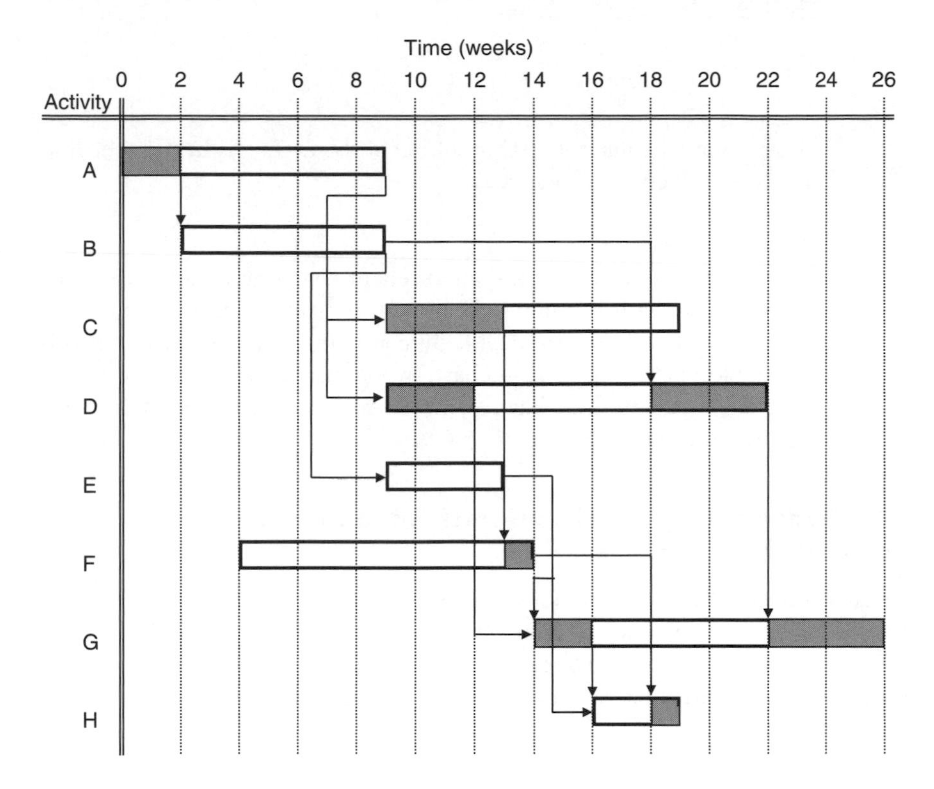

Figure 5.9 A linked bar chart of the overlapped precedence schedule.

The bar chart in Figure 5.9 reveals that activity F has no preceding link and should therefore be regarded as another start activity. Since the schedule now has two start activities, A and F, for clarity a dummy activity, 'START', is inserted into the schedule (see Figure 5.8). The value of lag START–F was calculated as 4.

Step 5: Calculate values of LSDs and LFDs

LSD and LFD values of the activities in the schedule can easily be calculated from the expression of total float given in Chapter 3 as:

$$TF_I = LFD_I - EFD_I$$

By rearranging the formula, the LFD of activity I is then:

$$LFD_I = EFD_I + TF_I$$

The LDS of activity I is then:

$LSD_I = LFD_I - Duration_I$.

The calculated values of LSD and LFD of the activities in the schedule are given in Figures 5.7 and 5.8.

Step 6: Plot a linked bar chart
While a precedence network is an excellent computational tool, it lacks the ability to clearly communicate the planning information, particularly when the network is overlapped. By converting a precedence network into the format of a linked bar chart, overlapped links between activities are clearly apparent (see Figure 5.9). Identification of all start and end activities is then a relatively simple task.

5.9 Overlapping of critical path schedules by computer

Most commercially available CPM software contains built-in overlapping functions that allow the planner to construct realistic schedules. Some software, such as Primavera P6, provides the full range of overlapping models. Overlapping of critical path schedules using computers will be examined in Chapter 7.

5.10 Redundant links in precedence schedules

When three sequential activities, A, B and C, in a precedence schedule with only FTS links form a triangle, the link AC is redundant (see Figure 5.10(a)). This is because the relationship between activities A and C has already been established by the link ABC.

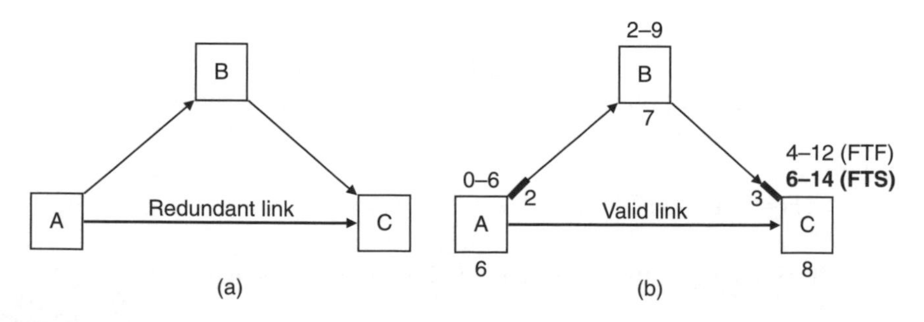

Figure 5.10 A redundant and a valid link in a precedence schedule.

However, when a precedence network is overlapped, the link AC may or may not be redundant. Let's examine a simple precedence network in Figure 5.10(b) that contains some overlaps and check whether the link AC is redundant. To determine this requires calculation of the schedule and determination of the ESD value of activity C. When the schedule is calculated, the ESD of activity C is either 4 (for the FTF link between B and C) or 6 (for the FTS link between A and C). By definition, the ESD of activity C must be 6 (the greater of those two values of ESD). It means that activity C cannot start until activity A is fully completed. The link AC is therefore valid and must be included in the schedule.

5.11 Summary

This chapter has introduced the concept of overlaps for use in critical path scheduling. A number of standard overlapping models were examined in detail and their implications illustrated on an example. By applying the defined overlapping models such as STS, FTF, STF and the compound overlap involving both STS and FTF, the planner is able to realistically schedule the production process.

The next chapter will examine the use of the CPM as a monitoring and control tool.

Exercises

Solutions to the following exercises can be found on the following website: http://www.routledge.com/books/details/9780415601696/

Exercise 5.1

An overlapped precedence schedule is given in Figure 5.11. Calculate the schedule and determine:

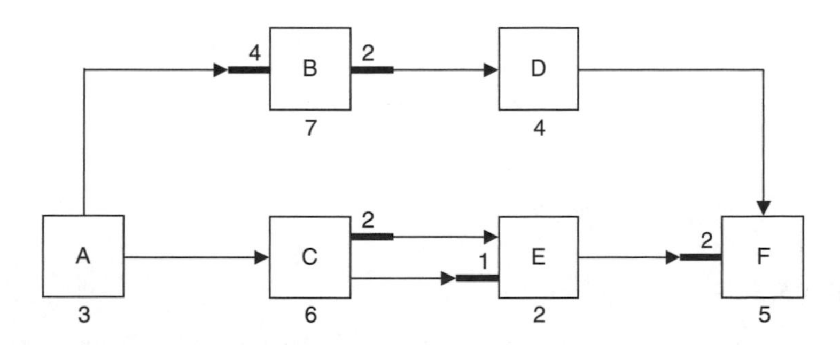

Figure 5.11 The precedence schedule.

1 The ESDs and EFDs of all activities in the schedule
2 Lags of all links in the schedule
3 The critical path
4 Free and total float of non-critical activities
5 The LSDs and LFDs of all activities in the schedule.

Convert the calculated precedence schedule to a linked bar chart.

Exercise 5.2

A portion of an overlapped precedence schedule is given in Figure 5.12. Calculate the schedule and determine:

1 The ESD and EFD of activity O
2 Lags of all links in the schedule
3 Free and total float of all activities in the schedule
4 The LSD and LFD of activities K, L, M, N and O.

Identify the true terminal activity.

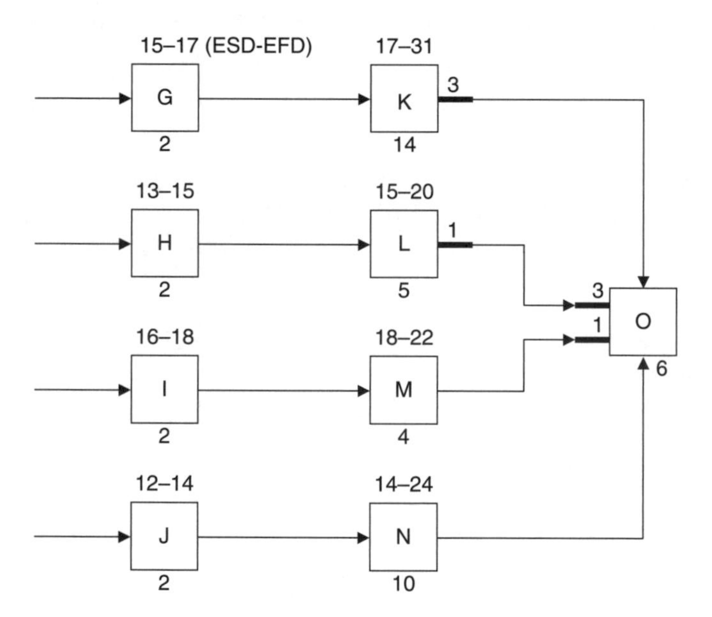

Figure 5.12 The precedence schedule.

Exercise 5.3

An overlapped precedence schedule is given in Figure 5.13. Calculate the schedule and determine:

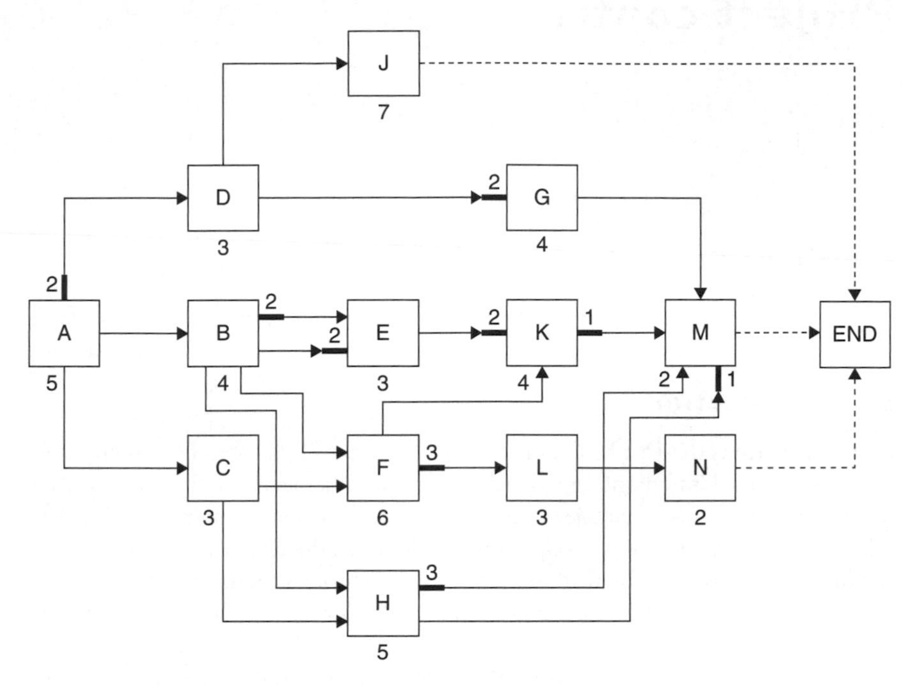

Figure 5.13 The precedence schedule.

1 The ESDs and EFDs of all activities in the schedule
2 Lags of all links in the schedule
3 The critical path
4 Free and total float of non-critical activities
5 The LSDs and LFDs of all activities in the schedule.

Convert the calculated schedule to a linked bar chart.

Chapter 6

Project control

6.1 Introduction

Project control is the third element in the functional chain of management after work has been planned and organised. It completes a closed circuit or loop that will enable the project manager to detect deviations from the plan that could be related to organisational or production problems, inefficient use of resources, or the uncertain nature of the work in question. The project manager will be able to take corrective action by adjusting the production process and/or improving the use of resources.

In this chapter project control will be defined from the scheduling point of view and its importance explained. The emphasis will be on time and resource control and cost–time optimisation. A method for compressing or accelerating critical path schedules will be described and demonstrated. The technique of earned value, which is an effective method of controlling the progress and performance of projects, will be discussed later in the chapter.

The main function of project control is to ensure successful attainment of the objectives that have been formulated in plans. Since plans are forecasts of future events, they are never truly accurate and the project manager must expect that deviations from the planned strategy will occur in the production process. An effective project control system will be able to detect any such deviations and will enable the project manager to formulate and implement remedial strategies.

6.2 Project performance outcomes

Project objectives defined in the planning stage impose specific performance outcomes that a project is expected to attain. These are usually related to time, cost and quality performance. Project control thus aims at defining performance outcomes and setting up a mechanism for their achievement.

Depending on the type and nature of a project, a control system may include additional performance factors such as control of change orders, production of documentation, and control of safety. Time progress control and resource productivity control factors will be the main focus of discussion in this chapter.

6.3 Project control system

The project control system ensures that the work meets the defined performance outcomes. It requires that the plans accurately replicate strategies developed for the execution of the project in accordance with the design documentation. It is able both to detect deviations from the plans that can be evaluated and to take periodic corrective action to bring the work into line with the plans (Harris 1978).

The type and complexity of a control system depends on the nature of a project and its size. A small project such as construction of a cottage would probably require no more than a simple bar chart schedule and a simple cost-reporting mechanism. Control of large projects, on the other hand, requires the development of a more complex system for measuring performance of a range of important factors. The main tools of such a control system are:

- A target schedule
- A cost budget based on the master cost estimate or master cost plan
- A quality assurance plan
- A reporting mechanism for the use of resources and other tasks.

A resource-levelled target schedule is an important tool of cost control. It determines a norm against which deviations in time and in the use of resources can be measured. The fundamental requirement is that a target schedule accurately represents the work that is to be accomplished in the time specified. The target schedule is often referred to as a baseline schedule in planning software such as Primavera P6. Chapter 7 illustrates an example of a baseline schedule.

A cost budget, which is based on the master cost estimate or master cost plan, is the main tool of cost control. Apart from assembling information on costs incurred to date, it reports deviations from the master estimate and forecasts future expenditures. It also provides a basis for cash flow control.

A quality assurance plan defines the required quality performance outcomes and provides a mechanism for measuring them. For more information, see Oakland and Porter (1994), Dawson and Palmer (1995), Gilmour and Hunt (1995) and Oakland and Sohal (1996). Examples of reporting mechanisms of use of resources will be presented later in this chapter.

The project control system involves monitoring, evaluation and adjusting or updating. These three important components of project control are graphically shown in Figure 1.1 in Chapter 1. A control system should be designed and implemented as a continuous process. It should be activated immediately after the start of production and should be maintained for the entire period of the project. Following the completion of the project, the project reporting information and schedule data should be captured and recorded in the form of a database. The database is useful in estimating activity durations of similar future projects. Activity estimate techniques and databases are explained in Chapter 11. The least expensive solution of a problem is its timely resolution. It means that the control system must be capable of reporting feedback information regularly and on time. This can be achieved through the application of a computer-based system.

6.4 Monitoring performance

Monitoring generates feedback about progress that has been achieved with regard to time, resources, costs, quality and other aspects of the project. Monitoring of time and resources will now be discussed in more detail.

6.4.1 Monitoring time performance

Time performance of projects may be monitored visually, by direct communication with production personnel or by formal meetings.

The visual assessment of progress assists the project manager not only to establish the time performance of resources but also in becoming aware of emerging problems, which may directly or indirectly affect progress. For example, the project manager may note an increasing volume of accumulated waste, the presence of bottlenecks, the inadequacy of storage facilities, or even inconsistency in the quality of finishes. Feedback on time performance may also be obtained through informal communication with production personnel. Regular site meetings provide the main forum for monitoring progress of work. The status of the project is determined from the schedule on which the actual progress achieved has been marked.

The time performance of large projects can also be monitored by focusing on one or a small number of key activities, which often tend to be the critical resources of the critical path. For example, progress of a high-rise commercial building designed as a concrete frame is largely governed by the speed of the formwork activity. If the speed of forming columns, walls, beams and slabs slows down, so will the speed of other succeeding activities, the result of which may be delay in completing the project. In this case, it would be appropriate to determine from a project schedule the planned volume of formwork to be installed each day or each week. Monitoring would then compare against the plan the actual volume of formwork completed. It makes sense to include monitoring of performance of the key activities within the overall monitoring strategy.

6.4.2 Monitoring the use of resources

Feedback on the use of resources such as labour, materials and plant/equipment involves regular collection and recording of relevant information. It is essential that such information is generated accurately and on time.

Feedback information on the use of direct labour is generated by the timesheets that a supervisor compiles for each week; from these, the labour cost is calculated. Time-sheets include highly detailed information on what each worker did during the monitoring period. Against each task performed by a worker the supervisor records an appropriate cost code. Every effort must be made to ensure completeness and accuracy, which is essential for accurate reporting of costs. An example of a time-sheet is given in Figure 6.1. With the advent of computer generated scheduling, time-sheets are often created and completed electronically. Their integration into the project schedule assists in streamlining the monitoring process. More information on electronic time-sheets can be found in Chapter 7.

Since labour resources are committed to activities in the schedule, it is essential to monitor how these resources are being used. Time-sheets may help in revealing excessive overtime or periods of idle time. They also provide information on the total volume of the workforce working on the project each day.

A project schedule also helps to identify potential areas of inefficiency in the workforce by showing activities that have been delayed. The project manager is then able to determine whether or not such delays are caused by the workers.

PROJECT NAME:.....................................

Name of employee:................................ Week ending:............................

Employee no.:...................................... Supervisor: ....

Employee category:................................ Date:....................................

Working hours	Work codes						
	Monday	Tuesday	Wednesday	Thursday	Friday	Saturday	Sunday
7–8							
8–9							
9–10							
10–11							
11–12							
12–13							
13–14							
14–15							
15–16							
16–17							
17–18							
18–19							
19–20							

Total normal working hours:............

Total overtime hours:....................

Figure 6.1 Example of a time-sheet.

Monitoring the use of materials involves keeping a record of both delivered materials and materials in stock. The quantity of delivered materials is detailed on delivery dockets that the supervisor or another authorised site person is required to verify. The supervisor assigns an appropriate cost code to each delivered item and this is detailed on a delivery docket for cost control purposes. The type and quantity of delivered materials is then recorded in a stock control ledger, which has entries for receipts and issues of materials.

Monitoring the usage of plant/equipment requires the generation of a stock card database system and weekly plant/equipment time-sheets. A 'stock card' is prepared for each piece of plant/equipment that has been purchased or hired, showing the type, make and model, when purchased or hired, the name and details of the supplier, when due for service or return, and any other relevant information. With hired plant/equipment, the database should remind the supervisor of the impending due date for return. 'Weekly plant/equipment time-sheets' show the extent of use of each piece of plant/equipment that has been committed to the project. When cost-coded, this information becomes a part of a cost-reporting system.

The project manager may also rely on a schedule to monitor the efficiency of committed plant or equipment by assessing causes of delays and determining whether or not they may be attributed to the state of the plant or equipment.

6.5 Evaluating performance

The purpose of evaluation is to analyse feedback data collected on performance in a particular period. The process of evaluation establishes what deviations, if any, have occurred and if they have occurred, what caused them and what strategies are needed to bring the project back to its predetermined level of performance.

6.5.1 *Time performance evaluation*

The feedback data on time performance reveals whether or not the project is on schedule and the status of each activity and contingency in the schedule. The project manager is able to determine what activities have been delayed and what caused their delay, what activities have been fully completed and what activities are in progress. For those activities that are in progress and for those not yet begun, the project manager may, based on the progress feedback, reassess their duration and adjust allocation of resources accordingly.

Evaluation of performance of key project activities, such as formwork, is best carried out by a trend graph. A trend graph is an X–Y chart that shows performance (in cumulative terms) in relation to a time-scale. As the name suggests, a trend graph shows visually the development of trends in performance, which may be positive or negative. The important thing to remember is that once a specific trend in performance develops, it usually continues unabated until

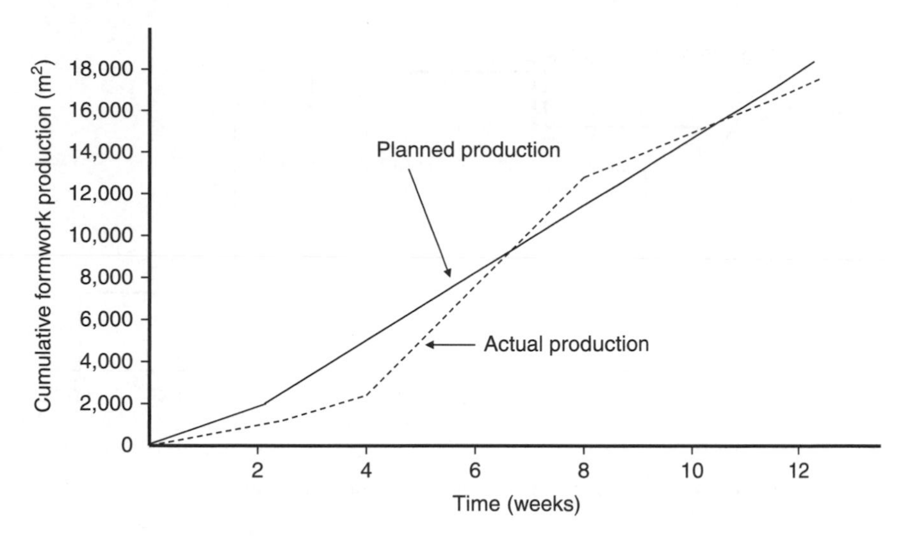

Figure 6.2 Example of the trend graph for cumulative formwork production.

management action alters it. This is illustrated in Figure 6.2, which shows the development of a negative trend of the actual formwork production in the first four weeks and the change in the trend caused by management action on week 4. The trend graph also shows another management action that was taken on week 8.

6.5.2 Resource performance evaluation

Apart from providing information for cost reporting, time-sheets of labour input also show the volume of overtime incurred or unproductive or idle periods. This may be indicative of insufficient or excessive volume of the labour resource or some other problem. Time-sheets also help to determine the total volume of the workforce employed on the project each day, which can be plotted as a histogram (see Figure 4.11). This information is crucial for reviewing the effectiveness of the personnel handling equipment, the volume of amenities, safety equipment on the site, and the adequacy of supervision. Since time-sheets provide data on the volume of incurred labour hours, this information can be plotted in a trend graph format together with the planned labour demand (see Figure 4.12).

Delays in the project schedule may indicate inadequate allocation of the labour resource in some activities. The project manager would attempt to alleviate this problem through resource levelling. Alternatively, the project manager may study the distribution of resources and the flow of work using a multiple activity chart. This topic will be discussed in Chapter 9.

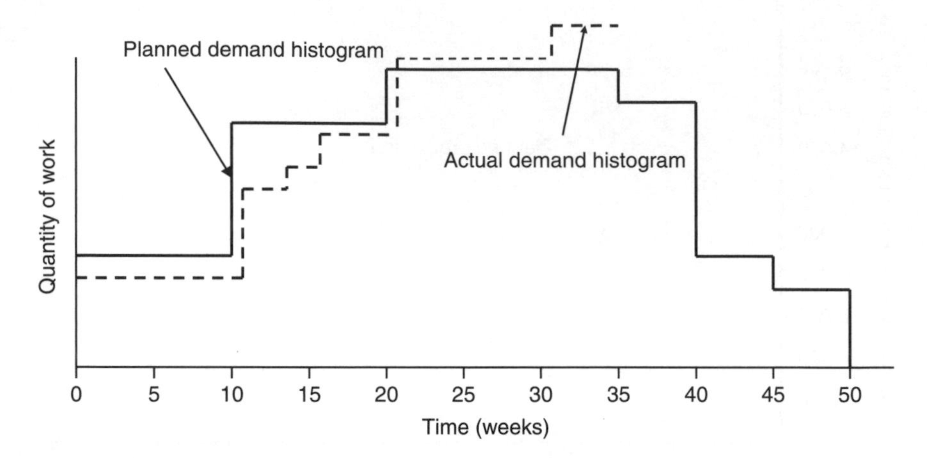

Figure 6.3 Histograms of the planned and actual demand of a crane.

Performance evaluation of materials involves comparing incurred costs with the planned cost estimate. Deviations may point to errors in quantities, wastage, or even the possibility of pilferage.

The comparison of incurred costs and the planned cost estimate is also applied to evaluating the performance of plant or equipment. Comparing histograms of the planned and achieved resource demands is another way of evaluating such performance. This is illustrated in Figure 6.3.

The outcome of evaluation is the formulation of specific strategies or decisions aimed at bringing the project back on track. This could require a change to the design, which may lead to modifications of the production process itself, reallocation of resources or injection of more resources. The cost and time impact of these actions is directly related to the magnitude of deviations between the planned and actual performance uncovered in monitoring.

6.6 Schedule adjustments/updates

With all the feedback data in hand and decisions made on how best to improve performance of a project, the project manager proceeds to adjust or update production schedules, cost budgets, and other control charts, graphs and histograms. Where possible, the project manager will use computer-generated tools for planning and control such as a critical path schedule and a cost budget; these make updating much easier and they can facilitate control cycles more often than manually generated planning tools.

Using two examples, a simple schedule will firstly be updated and recalculated based on its progress to date. A status report will then be prepared to the 'project control group' to assist with strategic decision-making.

6.6.1 Example of adjusting or updating of a critical path schedule

Computer-generated critical path schedules are widely used for both planning and control of construction projects. Computers are, however, of little help in extracting feedback data on the project's progress and evaluation of that progress. These are the tasks that are performed manually, even though they are laborious and time-consuming. The main benefit of computers is in performing the updating or recalculating phase of the control cycle, which they can do almost instantaneously. Updating of a computer-generated critical path schedule will briefly be discussed in Chapter 7.

The aim of this section is to gain better understanding of the process of monitoring, evaluation and adjustment/updating through its application to a simple, manually generated critical path schedule. The project in question is office and showroom fitout of a small business enterprise. A precedence schedule of the project prior to the next progress review scheduled for week 7 is given in Figure 6.4.

A project control system operates on weekly cycles. It is now week 7 of the project period. A new control cycle has just begun. The process of monitoring established a status of progress of each activity in the schedule in terms of whether or not the activity has been completed, not yet started or is in progress (Table 6.1).

Evaluation of the performance of individual activities revealed, among other things, two serious problems. A substantial time overrun has occurred in the activity 'Showroom fitout', which was caused by the delay in obtaining the customs clearance for imported materials; and the activity 'Carpet' will now take significantly more time to complete because the floors have to be

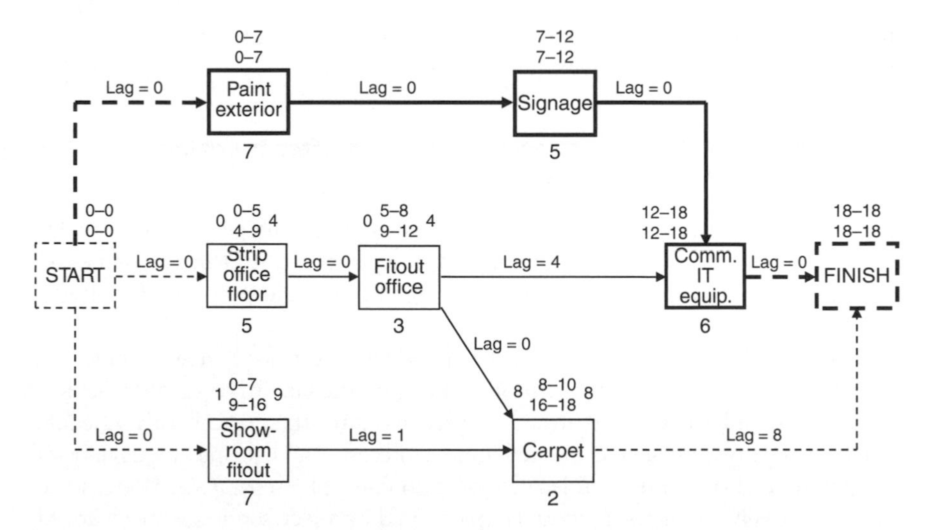

Figure 6.4 A precedence schedule of the fitout project prior to the progress review.

Table 6.1 The feedback data for the fitout project

Activities completed	*Activities not started (duration revised)*	*Activities in progress (duration to complete)*
Paint exterior	Signage (5)	Showroom fitout (5)
Strip office floor	Carpet (9)	
Office fitout	Commission IT equipment (6)	

decontaminated before the carpet is installed. Consequently, the activity 'Showroom fitout' is estimated to take a further five weeks to complete. The activity 'Carpet', which has not yet started, is now estimated to take nine weeks to complete rather than the original estimate of two. Durations of the remaining activities that have not yet started remain unchanged. Table 6.1 provides information on the revised durations of the activities, which are shown as a number in brackets.

The project schedule can now be updated. First, a line is drawn through the schedule to highlight the date of review, which is week 7. Because the precedence schedule is not drawn to scale, the line will not cut the schedule vertically. Rather, it will be drawn so that the completed activities are located to the left of the line and those that have not yet started to the right; those in progress will have the line passing vertically through them, as shown in Figure 6.5. Second, durations of the activities in the schedule are adjusted in accordance with information in Table 6.1. The schedule is now ready to be recalculated from the line that represents the date of review on week 7. The recalculated schedule is given in Figure 6.5. The outcome of the review on week 7 is the shift in the critical path and the extension of the project duration from 18 to 21 weeks.

6.6.2 Example of a status report produced after a critical path schedule has been updated

Once the critical path schedule has been recalculated, time performance evaluated and other control documents updated, the project status can be re-forecast and communicated to the 'project control group' for the purpose of strategic decision-making.

Figure 6.6 shows an example 'Weekly program report' for a new commercial office building that requires the existing building on the land to be vacated, demolished and re-built. The report would be prepared after the critical path schedule had been time performance evaluated and re-forecast, highlighting any deviations and potential effects on the project completion date. The separate sections of the report effectively summarise potential impacts on the project, serving as an advanced warning tool and enabling management to control the direction of the project.

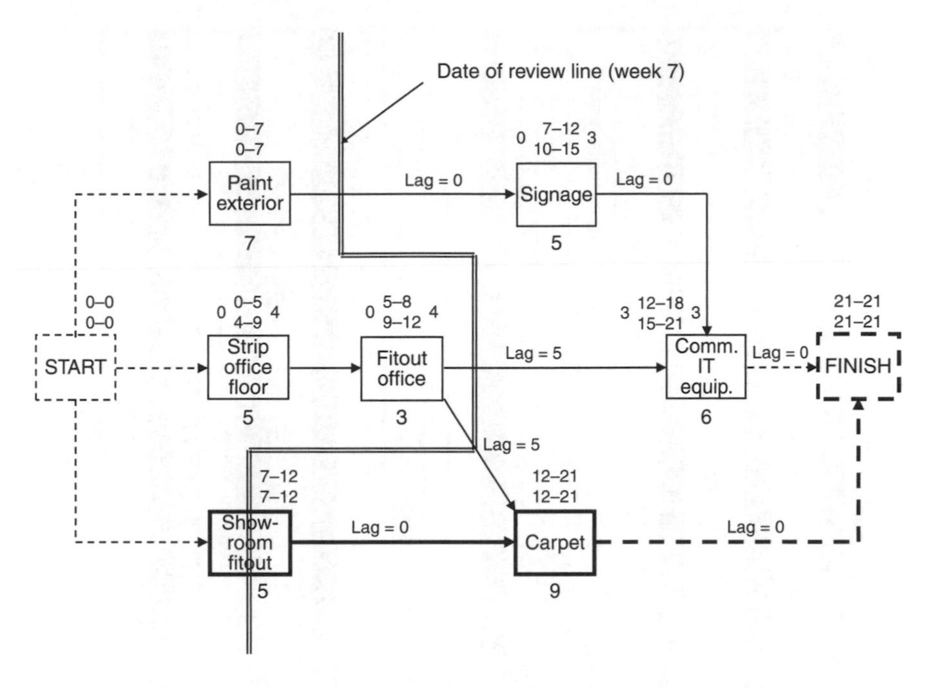

Figure 6.5 A precedence schedule of the fitout project after update on week 7.

For 'program reporting' to be effective, it is best undertaken no less than every fortnight with all key project stakeholders involved. The program report can be adapted to suit the project specifics.

6.7 Cost–time optimisation

The typical characteristic of construction projects is that they are often completed late. There are many reasons for this, such as design changes, lack of planning and inappropriate use of planning tools, adverse climatic conditions, latent site conditions and often simply unrealistic expectations.

The delay in completion of a project is usually accompanied by an increase in its cost due to the imposition of liquidated or delay damages. If the delay and the extra cost are the responsibility of the client, the main contractor is entitled to a claim under the contract for extension of time and for costs. If, however, the contractor is responsible for the delay in completing the project, the contractor may be liable to pay crippling liquidated damages to the client, in addition to being required to absorb any extra costs. The contractor would naturally attempt to make up for the delay, which is punishable under the contract by shortening the project period (see Uher and Davenport (2009) for more information).

SAMPLE WEEKLY PROGRAMME REPORT

Print Date: 19-Dec-10

CLIENT/BUILDER CO-ORDINATION

Events		Start Time	Finish	Co-Ordination Requirements
Client vacate existing building managers offices and cleaning rooms.		17-Jan-11	24-Jan-11	All loose furniture and equipment to be removed by Client. Builder to isolate and disconnect services

NOTIFACATIONS

Stakeholder	Works	Impact	Works Commence	Action
Existing building level 3 tenant	Tenant to vacate building.	Major	~~08-Dec-10~~	Notice to vacate issued. Tenant to defit and vacate.

KEY MILESTONES STATUS

ID No	Level	Activity Description	Contract Start Date	Contract Completion Date	Start Date	Completion Date	This Weeks Status	Last Week Status	Change in week	Assessment/Actions
					Forecast	Forecast				
6	ALL	Existing building fully vacant for demolition		24-Jan-11		31-Jan-11	- 1 week	0	- 1 week	1 week late due to asbestos discovered. Remediation underway.
48	ALL	Construction Works Commence	27-Jan-11		07-Feb-11					
2003	1-5	Tenant 1 Area Complete		16-Mar-12		16-Mar-12				
2018	6-10	Tenant 2 Area Complete		04-May-12		04-May-12				
2018	6-10	Practical Completion		15-Jun-12		15-Jun-12				

CONTRACT LETTING

Contact Package	Status	Planned Let By'	Fc'st Let By'	Notes
Site Setup (Fences, hoardings, accom)	MTS	01-Feb-11	01-Feb-11	
Surveyor	Tender Close	01-Feb-11	07-Feb-11	Potential delay in surveyor start on site. Only two contractors have submitted tenders on time, Project Manager to call Head Office.
Traffic Management	Tender Close	01-Feb-11	16-Feb-11	
Civil Contractor (Bulk, detailed, topsoil)	Design Due	08-Feb-11	20-Feb-11	
Hydraulic Services	Design Due	26-Apr-11	26-Apr-11	
Electrical Services	Design Due	10-May-11	10-May-11	

PROCUREMENT

Item	Order Date	Lead Time	Planned Start On Site	Forecast Start On Site	Notes
Lifts	15-Apr-11	140 calendar days	10-Aug-11	10-Aug-11	Design to commence by no later than 31-Jan-11

LIKELY AREA'S OF RISK/OPPORTUNITY

Item	Impact	Likeliness	Proposed Corrective Measures/Actions
Existing building tenant to vacate - Asbestos discovery	High	Likely	Builder to submit extension of time as asbestos is a latent condition. Acceleration options to be explored.
Surveyor Tender - 2 tender submissions only	Low	Highly Likely	Project Manager speak with Head Office to seek approval for 2 tender submissions.

KEY DECISIONS ACTIONS

Item	Comments
Surveyor Tender - Head Office approve 2 tenders.	Commercial Risk.

Figure 6.6 Example of a weekly program report.

The contractor may also want to complete the project ahead of the schedule in order to transfer the resources from the current project to a new one, which may be scheduled to start before the completion date of the current project. Another reason why the contractor would want to hasten completion is to earn a financial bonus from the client for completing the project ahead of schedule.

Accepting that some delays are likely to occur and/or that opportunities for time-saving may arise throughout the life of a project, it is necessary to deploy a process for effective management of delays or opportunities for time-saving. Such a process is referred to as 'compression' or 'acceleration'. Its main function is to shorten the project duration by injecting more resources at the minimum possible cost. The key principle of compression or acceleration is to find an optimum cost–time solution.

Before defining the process of compression or acceleration in detail, it is necessary to examine the relationship between cost and time.

6.7.1 Cost–time relationship

For any given activity, the cost of performing it is related to the period of its performance. The relationship between cost and time is indirect. This is because the time for performing the activity decreases as the cost related to the resource input increases. Figure 6.7 illustrates the relationship between cost and time for a particular activity, which is not only indirect but also non-linear, as shown by a curve between A and C. Point C (coordinates $T_C C_C$) in Figure 6.7 represents the minimum or the normal cost, and point A (coordinates $T_A C_A$) represents the minimum or the crash time.

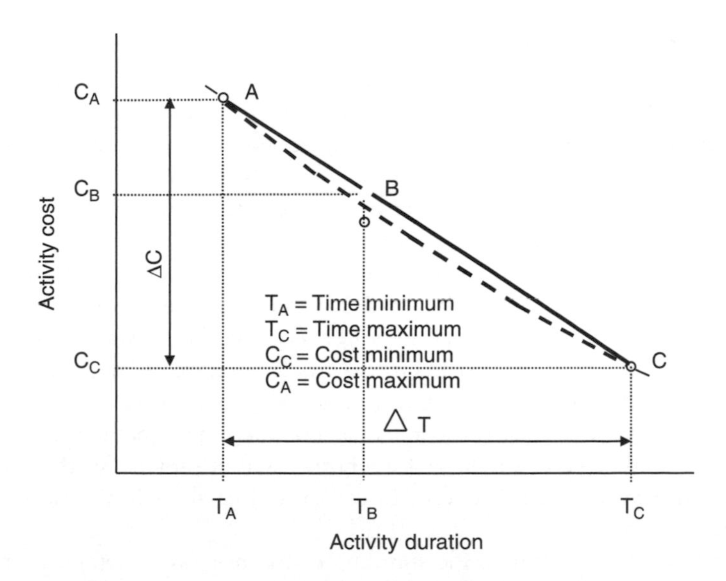

Figure 6.7 An activity's cost–time relationship.

It is possible without introducing much error to assume that the relationship between A and C is linear. Under this assumption, it is possible to express the value of the cost slope between the points of the normal cost, C, and the crash time, A, as:

$$\text{Cost slope} = \frac{\Delta C}{\Delta T}$$

where:

$$\Delta_C = C_A - C_C$$

$$\Delta_T = T_C - T_A$$

The assumption of linearity in the shape of the cost slope makes it possible to determine for a particular estimate of time the equivalent estimate of cost. For example, the duration T_B corresponds to the cost C_B.

As evident in Figure 6.7, if a project schedule is to be accelerated, more resources must be injected to perform the work in a shorter time. The faster the rate of acceleration, the greater the need for additional resources and the greater the increase in cost.

6.7.2 Cost components

The total project cost is composed of a number of component costs. For simplicity, only two of these will be considered in compression/acceleration. They are direct costs and indirect or overhead costs.

Direct costs are associated with the cost of labour (including on-costs), materials and plant/equipment. The cost of each project activity is estimated as direct cost. Together, direct costs of individual activities constitute the direct cost of the project. An interesting characteristic of the project direct cost is that its magnitude increases with the decrease in the project period. This is illustrated in Figure 6.8. However, when the project duration exceeds the normal cost point, the project direct cost may also increase. This is due to inefficiencies in the use of resources when the schedule is unnecessarily long.

Indirect costs are those associated with general overheads. They are costs other than direct costs. Indirect costs may be fixed or variable over a period of time. When the production output is steady, the indirect cost may be constant or fixed, but when the volume of production fluctuates, the indirect cost may increase with the increase in the production output. In this case, the indirect cost is variable. For example, if the contractor has only one construction project in progress, the contractor's overhead cost would be fully borne by that project. With more projects in hand, the contractor's overhead cost would be shared in some way between those projects.

For the purpose of developing understanding of the process of compression/acceleration that will be discussed in section 6.7.4, let's assume that indirect cost

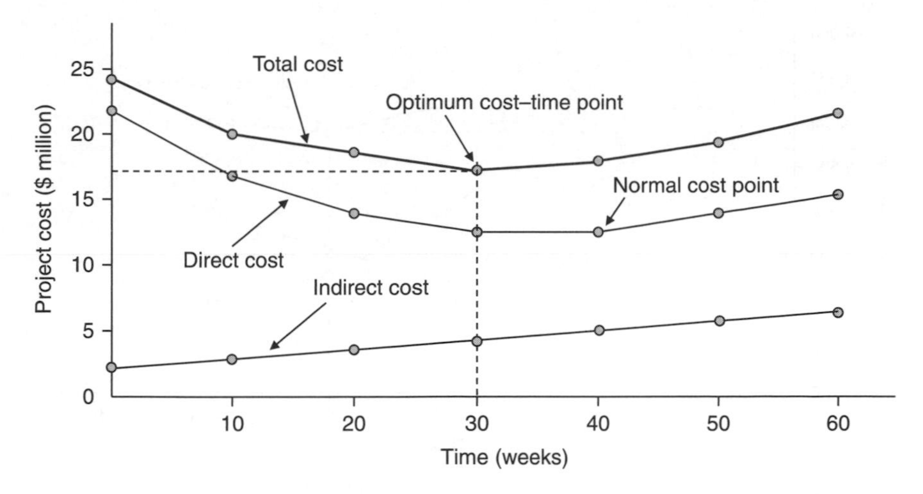

Figure 6.8 Project cost components.

is fixed over a period of time. In cumulative terms, the indirect cost is a rising cost over a project period. Graphically, it is expressed as a rising, straight line (see Figure 6.8). The total project cost is then the sum of the direct and indirect costs (Figure 6.8). The lowest point on the total cost curve represents the point at which the cost and the time are at optimum. In Figure 6.8 the optimum point on the total cost curve has the time coordinate of 30 weeks and the cost coordinate of $17.5 million.

6.7.3 Cost–time optimisation of activities

The concept of cost–time optimisation can be applied to find an optimum relationship between the duration of an activity and its cost. Let's consider a simple example in Table 6.2 in which one specific activity, say 'Excavate', could be accomplished in four different periods depending on the type and volume of resources allocated to it. Direct costs have been estimated for each period of work. The indirect cost is assumed to be fixed.

Table 6.2 Estimates of cost and time values for the activity 'Excavate'

Possible activity duration (days)	*Direct cost of activity ($)*	*Indirect cost/day ($)*
10	1,000	200
8	1,300	200
6	2,000	200
4	3,000	200

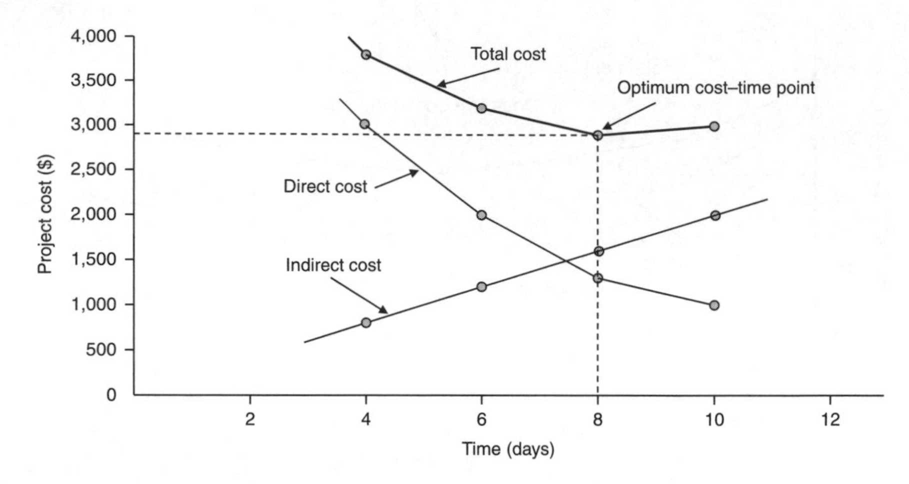

Figure 6.9 The optimum cost–time point of the activity 'Excavate'.

The analysis of this example is illustrated in Figure 6.9. The lowest point on the total cost curve represents the activity's optimum cost–time point. Its coordinates are eight days' duration and $2,800 cost.

6.7.4 Compression or acceleration analysis

The preceding sections of this chapter have examined the relationship between project cost and project time. This cost–time relationship forms the basis of the concept of compression or acceleration. It should be noted that the terms 'compression' and 'acceleration' mean the same thing.

When it is necessary to shorten the project schedule, the concept of compression helps to reduce its duration at the least possible cost. Compression is based on four fundamental rules designed to keep cost and time at optimum:

1 Only critical activities may be compressed.
2 The critical activity with the lowest cost slope (that is, the cheapest critical activity) is compressed first.
3 The amount of compression of the critical activity must be smaller than or equal to the amount of total float of non-critical activities positioned on a path parallel to the critical path. If the amount of compression is equal to the amount of total float of such non-critical activities, those non-critical activities will become critical and they will form an additional critical path. If the amount of compression exceeds the amount of total float of such non-critical activities, the original critical path will be lost and those non-critical activities will become critical. If this occurs, the following Rule 4 will be contravened.

4 The original critical path and any other additional critical paths that have been created during the compression analysis must be retained throughout the analysis.

Rule 1 states the obvious: only critical activities are able to affect schedule reduction. Compressing non-critical activities will incur cost without any reduction in time.

Rule 2 stipulates that if there is more than one critical activity capable of being compressed, the one that costs the least amount of money to compress should be compressed first.

Rule 3 sets the limit of compression, and together with Rule 4 guards against the loss of the original critical path or any other critical paths that have been created during the compression analysis. The original critical path is the benchmark against which the compression analysis is carried out within the context of cost–time optimisation.

The compression process is repeated several times until all possible alternatives have been exhausted. It should be noted that not all critical activities are capable of compression. It is the task of the project manager to determine the compression capabilities of individual critical activities. As the compression analysis progresses and more non-critical activities become critical (according to Rule 3), such new critical activities will also be considered for further compression.

The compression procedure will now be demonstrated step by step on a practical example.

The problem

The contractor was awarded a contract to build a block of high-rise apartments in 43 weeks for the total contract sum of $9,310,000. The contract contains a liquidated damages clause of $20,000 per week. Since the client is a new developer with a strategy to invest in more construction projects over the next ten years, the contractor is aware of the need to perform well in order to secure more work from this client in the future.

The contractor's tender summary comprises:

Direct cost	$7,140,000
Indirect cost (fixed) @ $40,000 per week	$1,720,000
Profit margin	$450,000
Total tender sum	$9,310,000

The contractor's project manager has reviewed the tender documentation, particularly the cost estimate and the time schedule, and concluded that if the project is to be completed within the contract sum of $9,310,000, the duration would need to increase to at least 48 weeks. Apart from incurring extra overhead costs, the delay of five weeks would also invoke the liquidated damages clause. Furthermore, the contractor also needs to contemplate the impact that the delay in completing the project is likely to have on the future business relationship with the client. The revised project schedule is given in Figure 6.10.

The project manager has decided to assess the impact of compressing the schedule to its original duration of 43 weeks on the project cost. The project manager has been able to identify only five activities that are capable of being compressed and has determined the limits of compression, together with relevant compression costs. These are given in Table 6.3.

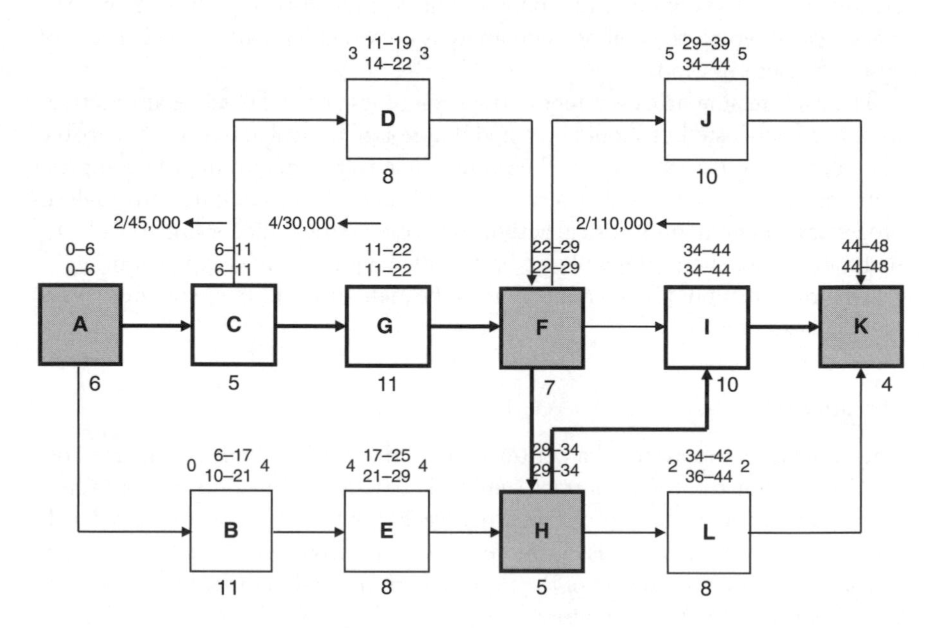

Figure 6.10 The project schedule before compression.

Table 6.3 Data for the compression example

Activity	*Normal time*	*Crash time*	*Cost of compression per week ($)*
B	11	9	35,000
C	5	3	45,000
D	8	6	40,000
G	11	7	30,000
I	10	8	110,000

Please note that the difference between 'normal time' and 'crash time' is the amount of time by which the activities in Table 6.3 can be compressed.

In Figure 6.10 and throughout the compression analysis, the following notations are used:

- When a critical activity in the schedule is shaded, it means that that activity is not capable of compression:

- The following notation indicates that activity G can be compressed by four weeks at $30,000 per week. The direction of the arrow is important here. When it points to the left, it indicates that four weeks are available for compression of activity G.

4/30,000 ←

G

11

- The following notation indicates that the activity G has been compressed by one week at $30,000 per week but may be compressed in the future by an additional three weeks. The fact that activity G has been compressed by one week is indicated by the arrow that points to the right. The other arrow, which points to the left, indicates that activity G has three weeks available for future compression. Because activity G has been compressed by one week, its duration is now ten weeks.

3/30,000 ← → 1/30,000

G

10

- The following notation indicates that activity G has been fully compressed by four weeks at $30,000 per week. All the available time for compression has now been exhausted. Duration of activity G is now seven weeks.

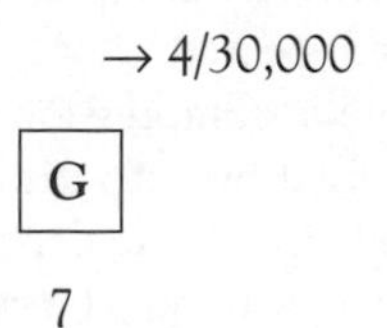

The project in Figure 6.10 will now be compressed by applying the four compression rules. Let's start with the first logical compression.

FIRST COMPRESSION

Compression Rule 1 states that only critical activities can be compressed. The critical activities in Figure 6.10 are A, C, G, F, H, I and K. Of those, only three are capable of being compressed (see Table 6.3). They are:

- Activity C by two weeks at $45,000 per week
- Activity G by four weeks at $30,000 per week
- Activity I by two weeks at $110,000 per week.

Compression Rule 2 states that the cheapest critical activity should be compressed first. The cheapest activity (per week) to compress is activity G. It costs $30,000 per week and may be compressed by up to four weeks.

Compression Rule 3 places the limit on the extent of compression, which is governed by the amount of total float of non-critical activities that are positioned on a path parallel to the critical path. Two non-critical paths lie parallel to the critical activity, G, that has already been selected for compression. They are CDF and ABEH, with the three and four weeks of total float respectively. It means that activity G can only be compressed by three weeks. When compressed, activity D will become critical and so will the path CDF. The values of total float of activities B and E will be reduced from four weeks to one week.

Therefore, activity G will be compressed in the first compression step by three weeks. The project schedule will now be recalculated to reflect the reduction in duration of activity G. The recalculated schedule is given in Figure 6.11. Its duration is now 45 weeks.

Project duration T_1 is 45 weeks

Cost of compression C_1 is $30,000 × 3 weeks = $90,000.

SECOND COMPRESSION

After the first compression step, activity D has become critical and since it can be compressed (see Table 6.3), it will be considered for compression together with the other critical activities in the second compression. The compression procedure defined in the first compression step will now be repeated.

The critical activities in Figure 6.11 that are capable of compression lie on two critical paths, the original one and the new one formed by activities CDF. To satisfy compression Rule 4, which requires the existing critical paths to be retained, the amount of compression along one critical path must be matched by

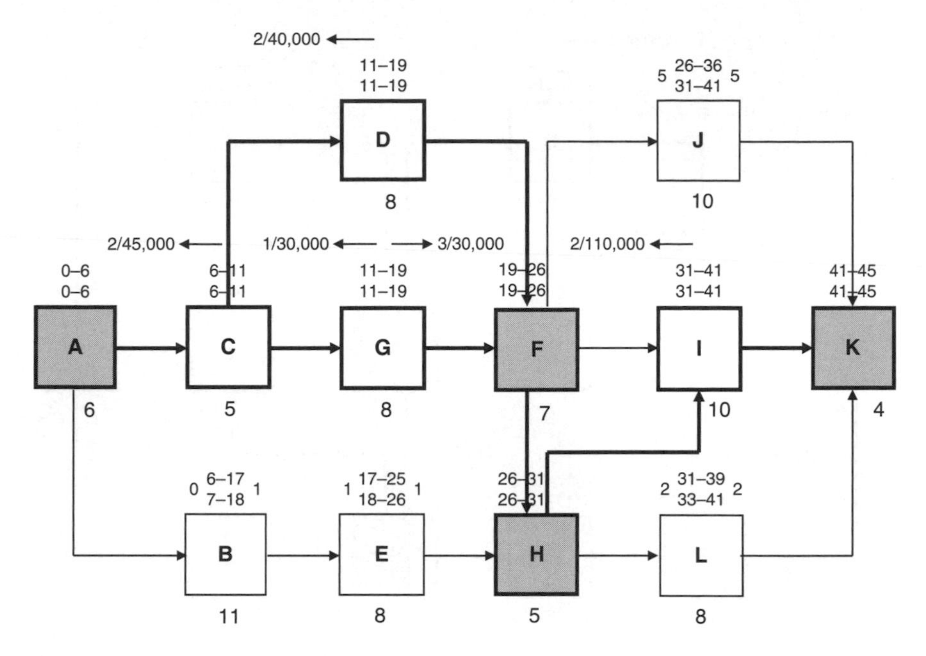

Figure 6.11 The project schedule after the first compression.

the same amount of compression along the other critical path. Let's identify critical activities that are capable of compression.

Since activity C lies on two critical paths (CDF and CGF), its compressing would shorten both critical paths by the same amount of time. Both critical paths would therefore be retained to satisfy Rule 4.

Neither activity D nor G can be compressed on their own since this would result in the loss of one of the two critical paths. Consequently, activities D and G would need to compressed together by the same amount of time. Activity I continues to be available for compression.

In summary, the following activities are available for compression:

- Activity C by two weeks at $45,000 per week
- Activities D and G together by one week at $70,000 per week
- Activity I by two weeks at $110,000 per week.

The cheapest activity (per week) to compress is activity C. Its compression costs $45,000 per week. It has two weeks available for compression.

The non-critical path ABEH is parallel to the critical activity C. Compressing activity C will reduce the amount of total float of the non-critical activities B and E. Since the amount of total float is one week, activity C can only be compressed by one week. After compressing C, activities B and E will become critical. A path ABEH will also become critical. Duration of C is reduced to four weeks.

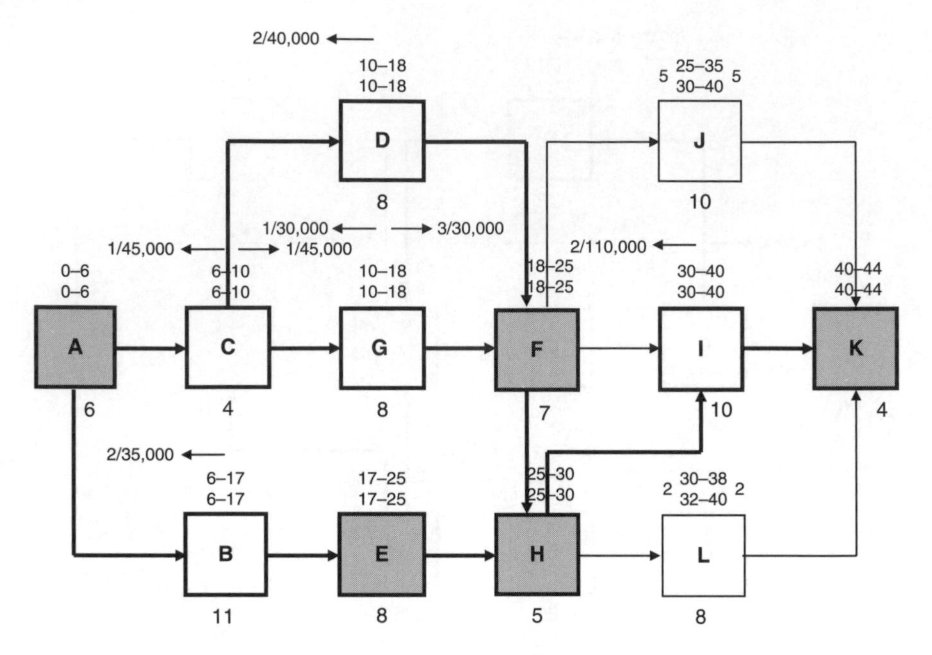

Figure 6.12 The project schedule after the second compression.

Therefore, activity C will be compressed in the second compression step by one week. The recalculated project schedule after the second compression is given in Figure 6.12. Its duration is now 44 weeks.

Project duration T_2 is 44 weeks

Cost of compression C_2 is \$45,000 × 1 week = \$45,000.

THIRD COMPRESSION

After the second compression, activities B and E have become critical, but only activity B is capable of compression (see Table 6.3). Only critical activities in the front part of the schedule are capable of compression. They are located on three parallel critical paths. The first path is ACDFH, the second is ACGFH, and the third is ABEH. It is important to apply compression Rule 4 here to ensure that all three existing critical paths in the front part of the schedule are retained after the third compression step.

Let's examine possible alternatives. Activity C cannot be compressed on its own because its compression would result in the loss of the critical paths ACDFH and ACGFH. In compliance with compression Rule 4, activity C would need to be compressed together with activity B.

Next, activities D and G are considered. If they are compressed together by the same amount of time in the third compression step, the critical paths ACDFH and ACGFH would be lost. In compliance with compression Rule 4, compressing activities D and G necessitates compression of B at the same time. The remaining alternative is to compress activity I.

In summary, the following activities are available for compression:

- Activities B and C by one week at $80,000 per week (the cheapest alternative)
- Activities B, D and G by one week at $105,000 per week
- Activity I by two weeks at $110,000 per week.

Therefore, activities B and C will be compressed together in the third compression step by one week. The recalculated project schedule after the third compression is given in Figure 6.13. The project duration is now 43 weeks.

Project duration T_3 is 43 weeks

Cost of compression C_3 is $80,000 × 1 week = $80,000.

The project schedule has now been compressed to meet the original contract period of 43 weeks. If required, the project schedule may further be compressed.

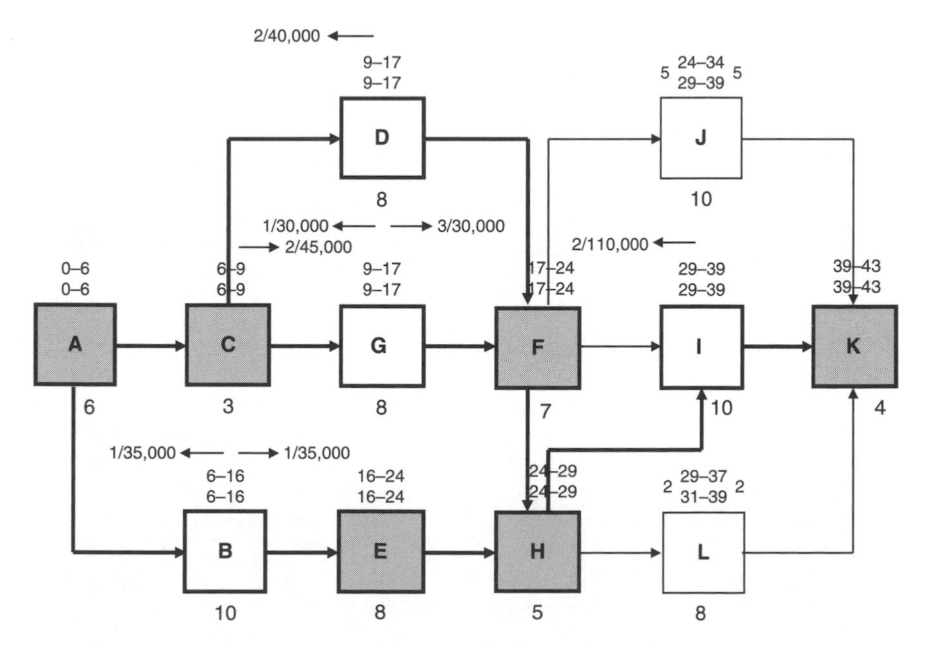

Figure 6.13 The project schedule after the third compression.

For example, in the fourth compression step, activities B, D and G could be compressed together by one week at $105,000 per week. In the fifth step, activity I could be compressed by two weeks at $110,000 per week.

The three compression steps that have brought the project schedule duration down to 43 weeks together with the starting scenario constitute four alternative solutions. Let's now calculate profit/loss outcomes for each of the four alternatives.

BEFORE COMPRESSION (T_n = 48 WEEKS)

Direct cost	7,140,000
Indirect cost 48 weeks × $40,000	1,920,000
Liquidated damages 5 weeks × $20,000	100,000
Total cost	9,160,000
Tender sum	9,310,000
Profit	*$150,000*

AFTER FIRST COMPRESSION (T_1 = 45 WEEKS)

Direct cost	7,230,000
Indirect cost 45 weeks × $40,000	1,800,000
Liquidated damages 2 weeks × $20,000	40,000
Total cost	9,070,000
Tender sum	9,310,000
Profit	*$240,000*

AFTER SECOND COMPRESSION (T_2 = 44 WEEKS)

Direct cost	7,275,000
Indirect cost 44 weeks × $40,000	1,760,000
Liquidated damages 1 week × $20,000	20,000
Total cost	9,055,000
Tender sum	9,310,000
Profit	*$255,000*

AFTER THIRD COMPRESSION (3 = 43 WEEKS)

Direct cost	7,355,000
Indirect cost 43 weeks × $40,000	1,720,000
Liquidated damages	0
Total cost	9,075,000
Tender sum	9,310,000
Profit	*$235,000*

Two important issues emerge from the compression analysis of the above project schedule:

1 Compression reduced total float of the non-critical activities and turned most of them into critical activities. Only two non-critical activities, J and L, remained in the project schedule after the third compression. The loss of total float in a project schedule reduces its capacity to absorb future delays. It also reduces its capacity to achieve the effective use of resources through resource levelling.
2 The first two compression steps actually improved the overall project profitability in comparison with the starting situation. However, profitability declined after the third compression. Beyond this point, the more the project schedule is compressed, the higher is the cost of compression and also the project cost. This is because activities that could be compressed in the later compression steps 4 and 5 cost more to compress. A higher cost of the latter compression steps is also attributed to the need to compress a group of critical activities.

With four possible alternative solutions in hand, what should the project manager do? There is no simple answer. In formulating an appropriate solution, the project manager would need to consider:

- The profit/loss outcomes of each compression step
- The likely damage to the contractor's reputation if the decision is taken to delay the contract
- The impact of the loss of total float and the increase in the number of critical activities on the ability to effectively manage the project
- The likelihood of securing time extension claims from the client that would, either partially or fully, offset the delay
- The current and future market conditions
- The level of construction activity in the future
- The project manager's own attitude to risk.

One plausible strategy that the project manager may consider is to inform the client of the problem and present the client with a range of alternative solutions including those developed through compression, together with an analysis of how such alternative solutions are likely to impact on both the client and the contractor. The client's response may help the project manager to formulate the most appropriate solution to this problem.

6.8 Earned value

The main objective of the control function, which is to measure the actual project performance and compare it with the plan, has already been discussed in this chapter. Time and cost are the two most commonly used performance measures, where time performance is assessed against a project schedule and cost performance against a cost budget. Assessing time performance independently from cost performance may provide misleading information about the overall project performance. This is illustrated in Figure 6.14.

The actual cost performance measured at a particular point in time, T_n, is 45 per cent of the cost budget, whereas it was expected to be 50 per cent. The project manager may conclude that the cost performance is better than expected. But unless the project manager relates this information to the time performance, it is not possible to establish whether the cost performance is good or bad.

The concept of 'earned value' was developed to assist the project manager in assessing time and cost performance in an integrative manner. Earned value represents a uniform unit of measure of progress in terms of time and cost, and provides a basis for consistent analysis of project performance. The technique of

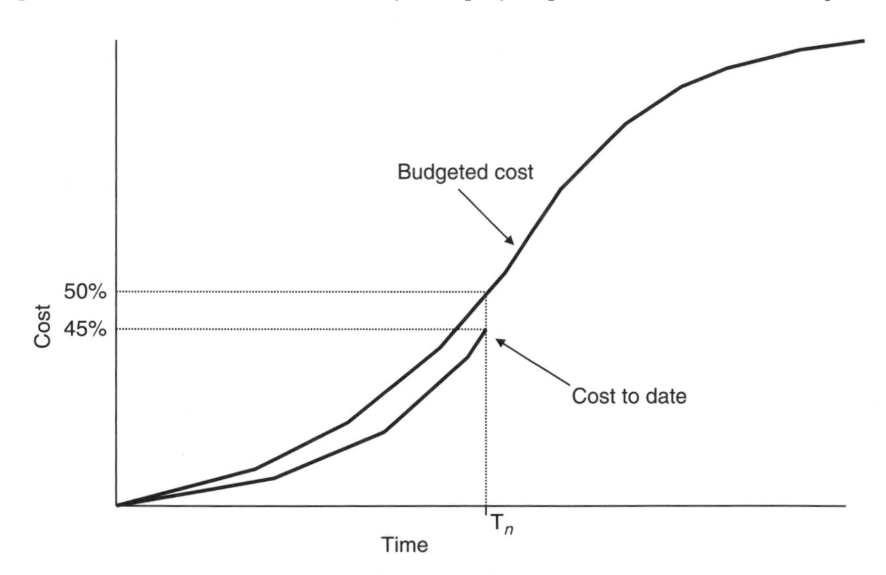

Figure 6.14 Assessment of project cost performance .

earned value was developed in the United States in the 1960s and has largely been used by the US Department of Defense. Since the 1980s, it has been extensively applied in defence projects in Australia.

6.8.1 The technique of earned value

The technique of earned value assists in assessing current progress and performance, and forecasting future progress and performance. To understand the concept of earned value, it is necessary to become familiar with the key elements of earned value that are defined by specific abbreviations and acronyms, which have been adopted worldwide. They are illustrated in Figure 6.15.

The first element is 'budgeted cost for work schedule' (BCWS). It is the project target cost and is expressed in Figure 6.15 as an S-curve. For a scheduled project duration Ts 'budget at completion' (BAC) is $10,000. This is the second element of earned value.

When a project gets under way, the control mechanism will generate feedback on progress at specific points in time known as 'data dates'. At a data date, the project manager determines in percentage terms the actual progress achieved. It is referred to as PC or 'percentage complete'. This is the fourth element of earned value. For

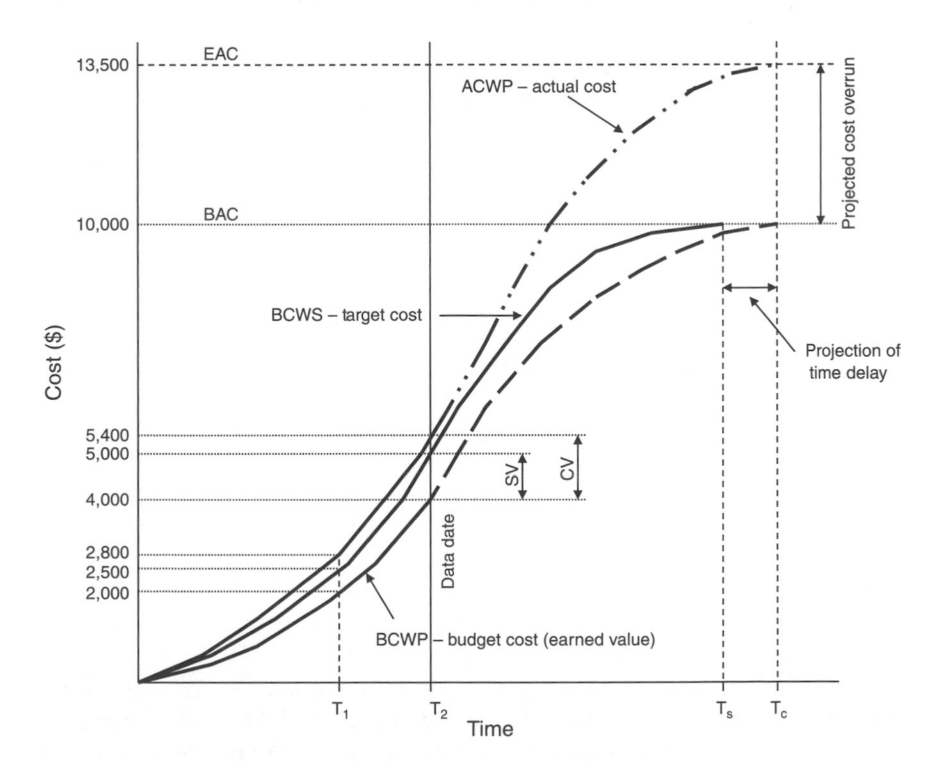

Figure 6.15 Earned value elements (adapted from Burke (1999: 205)).

example, at the data date T_1 in Figure 6.15, BCWS is \$2,500. Let's assume that the project manager assessed PC as 20 per cent. This information enables the project manager to calculate BCWP or 'budgeted cost for work performed' or simply 'earned value', which is the fifth element of earned value. For example, at the data date T_1:

$$BCWP_1 = PC_1 \times BAC = 20\% \times \$10\,000 = \$2{,}000$$

Similarly, at the data date T_2, where BCWS is \$5,000, PC was assessed, say, as 40 per cent. Then,

$$BCWP_2 = 40\% \times \$10{,}000 = \$4{,}000$$

The $BCWP_1$ and $BCWP_2$ values are the two points on the BCWP curve.

If the data date T_2 is the current date, the project manager will forecast BCWP for the remainder of the project. BCWP at completion will be equal to BAC, but the project period may be greater or smaller than T_s. In Figure 6.15 the project period for BCWP at completion is T_c.

While $BCWP_1$ at the data date T_1 is \$2,000, the actual cost incurred may be different. Let's assume that the actual cost is \$2,800. Similarly, assume that the actual cost is \$5,400 at the data date T_2, where $BCWP_2$ is \$4,000. These two actual costs are referred to as ACWP or 'actual cost for work performed'. They form an ACWP curve. ACWP is the sixth element of earned value.

In order to forecast ACWP at completion, the project manager calculates the value of EAC or 'estimate at completion', the seventh element of earned value, at the current data date. EAC is an estimate of the actual project cost at completion. It is calculated as follows:

$$EAC = (ACWP/BCWP) \times BAC \qquad (1)$$

Since

$$BCWP = PC \times BAC \qquad (2)$$

equation (1) can be simplified as follows:

$$EAC = (ACWP/PC \times BAC) \times BAC = ACWP/PC$$

At the data date T_2,

$$EAC = \$5{,}400/40\% = \$13{,}500$$

By extrapolation, the project manager is now able to draw an ACWP curve from the current data date to completion. By determining EAC and plotting an ACWP curve, the project manager is immediately alerted to potential time and cost overruns. This is probably the most important benefit of earned value.

When conducting periodic assessment of the project progress and performance, the earned value technique enables the project manager to determine, at specific data dates, schedule and cost variances, and schedule and cost performance indices.

The 'schedule variance', SV, the eighth element of earned value, is a difference between the earned progress BCWP and the budgeted progress BCWS. Although it is a measure of time variance, it is expressed in money units. The sign of the SV indicates whether the project is ahead of or behind the schedule. SV at the data date T_2 in Figure 6.15 is calculated as follows:

$$SV = BCWP - BCWS = \$4{,}000 - \$5{,}000 = -\$1{,}000$$

The negative sign indicates that the project is behind the schedule. The SV may also be expressed in percentage terms as follows:

$$SV\% = (SV/BCWS) \times 100\% = (-\$1{,}000/\$5{,}000) \times 100\% = -20\%$$

The 'cost variance', CV, the ninth element of earned value, is a difference between the earned value BCWP and the actual cost of the work. The sign of the CV again indicates whether the cost at a specific data date is greater or smaller than the original budget BAC. Let's calculate the CV as at the data date T_2.

$$CV = BCWP - ACWP = \$4{,}000 - \$5{,}400 = -\$1{,}400$$

The CV is negative, indicating that the cost is over the estimate. In percentage terms, the cost overrun is 35%.

$$CV\% = (CV/BCWP) \times 100 = (-\$1{,}400/\$4{,}000) \times 100 = -35\%$$

Schedule and cost performance indices represent another measure of progress and performance of a project. The 'schedule performance index', SPI, the tenth element of earned value, is the ratio between BCWP and BCWS. For the data date T_2 SPI is:

$$SPI = BCWP/BCWS = \$4{,}000/\$5{,}000 = 0.8$$

When the value of SPI is smaller than 1, as is the case above, the project is behind schedule.

The 'cost performance index', CPI, the eleventh element of earned value, is the ratio between BCWP and ACWP. For the data date T_2 CPI is:

$$CPI = BCWP/ACWP = \$4{,}000/\$5{,}400 = 0.74$$

The value of CPI above is smaller than 1, which indicates that the project spending is over the budget.

6.8.2 Benefits of earned value

In summary, earned value is a highly appropriate technique for assessing and forecasting progress and performance of projects. Its numerous calculations can easily be handled by computer software such as Primavera P6. Since the calculation of the estimate at completion, EAC is dependent on determining the value of percentage complete, or PC, the project manager needs to exercise care in correctly assessing values of PC. Where possible, the project manager should assess the work completed to date quantitatively, for example in terms of the number of bricks laid. However, when expressing PC subjectively, for example in terms of the 50–50 rule, the fundamental requirement for the project manager is to be consistent. According to the 50–50 rule, when the activity starts, it is assumed to be 50 per cent complete, and when it is accomplished, it is assumed to be 100 per cent complete.

The technique of earned value enhances the cost performance analysis of a project through the application of a consistent methodology and by providing a uniform unit of measure. The key elements of earned value form the terminology that has been universally adopted. A comprehensive bibliography on earned value can be found in Christensen (2002).

6.9 Summary

This chapter examined the function of project control. It defined the project control system in terms of monitoring, evaluation and adjustment. It then examined processes of monitoring and evaluation of time and resource performance. It also illustrated updating of a critical path schedule and examined the concept of cost–time optimisation and gave an example of a basic report for an updated critical path schedule. The application of cost–time optimisation in the form of compression was demonstrated on a practical example. Finally, this chapter described the technique of earned value and its benefits in controlling progress and performance of projects.

Exercises

Solutions to the following exercises can be found on the following website: http://www.routledge.com/books/details/9780415601696/

Exercise 6.1

A precedence schedule of a project is given in Figure 6.16. Compress the schedule in accordance with the data given in Table 6.4. The project indirect (overhead) cost is given as $200 per day.

1 Determine a profit/loss outcome for each compression stage
2 Determine the optimum project time and cost point.

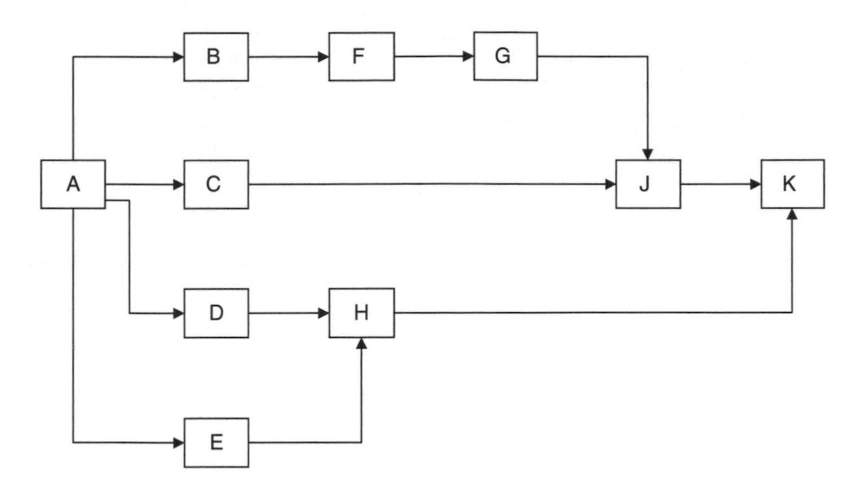

Figure 6.16 The precedence schedule for the example.

Table 6.4 Data for the compression example

Activity	*Normal time (days)*	*Normal cost ($)*	*Crash time (days)*	*Crash cost per day ($)*
A	3	600	2	200/day
B	1	50	–	
C	5	3,000	3	400/day
D	2	2,000	–	
E	4	1,200	2	100/day
F	1	300	–	
G	4	1,600	2	400/day
H	10	4,000	6	400/day
J	2	800	–	
K	4	2,200	3	700/day

Note: Read the table in the following manner: activity A can be compressed by one day from three days to two days, etc.

Exercise 6.2

A precedence schedule of a project is given in Figure 6.17. Compress the schedule in accordance with the data given in Table 6.5. The direct cost of the project is given as $10,000 and the indirect (overhead) cost as $100 per day.

1. Determine a profit/loss outcome for each compression stage
2. Determine the project cost when its duration is reduced to 17 days.

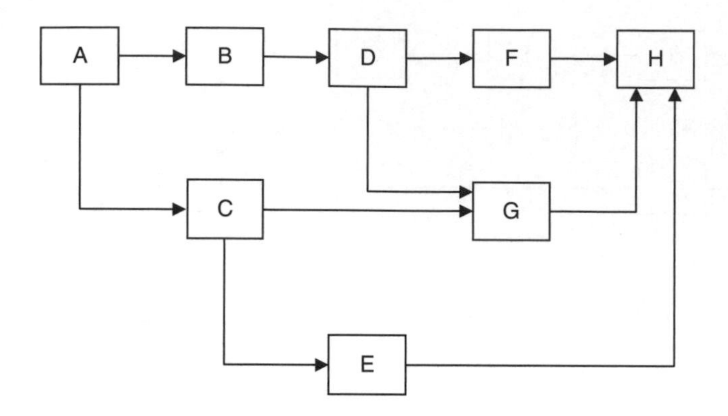

Figure 6.17 The precedence schedule for the example

Table 6.5 Data for the compression example

Activity	Normal time (days)	Crash time (days)	Crash cost per day ($)
A	3	2	50/day
B	2	1	150/day
C	7	4	100/day
D	5	4	200/day
E	4	3	50/day
F	6	4	100/day
G	2	2	
H	5	5	

Note: Read the table in the following manner: activity A can be compressed by one day from three days to two days, etc.

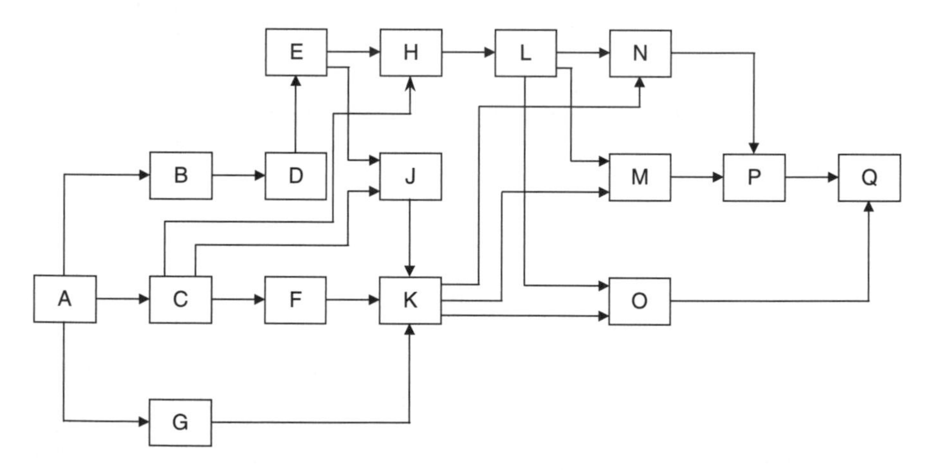

Figure 6.18 The precedence schedule for the example.

Table 6.6 Data for the compression example

Activity	*Normal time (days)*	*Crash time (days)*	*Crash cost per day ($)*
A	10	10	0
B	20	20	0
C	40	40	0
D	8	8	0
E	30	20	400/day
F	30	25	350/day
G	48	48	0
H	24	18	250/day
J	20	15	300/day
K	15	11	600/day
L	12	9	200/day
M	10	5	300/day
N	6	5	50/day
O	9	7	800/day
P	4	4	0
Q	4	4	0

Exercise 6.3

The contractor was awarded a contract for the construction of a small commercial building for the sum of $192,000. The contract period was agreed to be 108 days. The critical path schedule and the table of activities and costs for the project are given in Figure 6.18 and Table 6.6, respectively.

Before the start of the project,.the contractor reviewed the tender schedule and realised that the specified completion date could not be met. The revised schedule indicates that the project will take 122 days to complete (a delay of 14 days).

The contract conditions include the liquidated damages clause set at $100 per day.

The contractor's tender price comprised:

Direct cost	166,000
O/h cost 108 days × $200/day	21,600
Profit	4,400
Total sum	*$192,000*

Two weeks after the work on this contract began, the contractor was successful in securing another construction contract, but on a strict condition that the contractor would start the new contract before the completion of the current project (not later than on day 102 of the current project). The contractor expects to make $8,000 profit on the new project.

Analyse the problem from the contractor's point of view and evaluate a range of alternative responses.

Exercise 6.4

For a building project, the information in Table 6.7 was derived using the technique of earned value. For the given cases, calculate the values of EAC and interpret the project progress and performance (adapted from Burke (1999: 211)). Assume that the data date is approximately at a mid-point of the project period.

Table 6.7 Data for the earned value example

Cases	*BAC*	*BCWS*	*BCWP*	*ACWP*
1	$10,000	$5,000	$5,000	$5,000
2	$10,000	$5,000	$4,000	$4,000
3	$10,000	$5,000	$5,000	$4,000
4	$10,000	$5,000	$6,000	$4,000
5	$10,000	$5,000	$4,000	$5,000
6	$10,000	$5,000	$6,000	$5,000
7	$10,000	$5,000	$4,000	$6,000
8	$10,000	$5,000	$5,000	$6,000
9	$10,000	$5,000	$6,000	$6,000
10	$10,000	$5,000	$3,000	$4,000
11	$10,000	$5,000	$4,000	$3,000
12	$10,000	$5,000	$7,000	$6,000
13	$10,000	$5,000	$6,000	$7,000

Chapter 7

Critical path scheduling by computer

7.1 Introduction

The purpose of this chapter is to briefly examine computer-generated critical path scheduling. Primavera Project Planner PC software will be used to prepare a critical path schedule for a residential project.

As the popularity of electronic information systems has increased with the technology boom over the twenty-first century, computer-generated scheduling has virtually overtaken manually drafted schedules as the preferred method of critical path scheduling. The first generation of critical path method (CPM) software ran on large mainframe computers. The cost of computer processing was high and the speed of processing low, which restricted the use of CPM scheduling. The rapid advent of personal computers since the 1980s has reduced the cost of processing and, with the ever increasing speed of processing, revolutionised the use of CPM scheduling.

Today, there are many software products on the market to cater for a wide range of scheduling applications. They can generally be grouped into:

- Critical path scheduling software that performs high level scheduling, multi-functioned reporting and complex multi-computer network functions. Some examples of these software packages include Microsoft Project, Primavera SureTrak, Micro-Planner X-Pert, Primavera Project Planner P6 and Asta Powerproject.
- Critical chain scheduling software is available, but is generally an 'add-in' application package to the critical path scheduling software. Examples of these packages include Advanced Projects (MS Project Add-in) and ProChain Solutions.
- Line of balance software. Examples of these packages include Graphisoft and Misronet Q Scheduling.

Most CPM software products express duration of activities as a single value estimate. Some, including Primavera P6, Microsoft Project and Asta Powerproject, have the capacity to perform Monte Carlo simulation of a critical path schedule through add-in simulation software. Such software enables the planner to express duration of activities as a distribution of values that can be fitted to one of a number of standard probability distributions. The simulation software generates random numbers, which are then assigned to relative probability distributions of individual activities. The outcome of each iteration run is the calculation of one value of the project duration. Hundreds or thousands of iterations produce the final outcome in the form of the normal distribution of project time, which is fully described by its mean and standard deviation. The concept of probability, risk and simulation will be discussed in more detail in Chapter 12.

An alternative approach to probability scheduling is PERT (Program Evaluation and Review Technique). It will be examined in detail in Chapter 13.

This chapter will explore the use of Primavera Project Planner P6 software in scheduling a small residential project. It will demonstrate the generation of time and resource schedules, the use of overlaps, baseline scheduling and the control function and the cost-reporting facility. Because the focus of this chapter is on demonstrating the use of critical path software, the reader should not expect to find detailed information on how Primavera P6 works.

7.2 Brief overview of Primavera Project Planner P6 software

Primavera P6 is one of the leading CPM software packages, with a widespread use throughout the world. P6 is project, cost, and resource management software that allows organisations to create a collaborative environment through its set of integrated components and personalised interfaces for all project stakeholders. P6 improves on previous versions of Primavera, namely P3, by enhancing the Web-enabled technology and network-based databases (Oracle Corporation 2009).

The key modules that are included as parts of the Primavera P6 software package are:

1 'Project management', which includes the full range of functions necessary for planning and control of projects such as:

   i Time and resource scheduling
   ii Resource levelling for a large number of individual resources
   iii Modelling relationships through a full range of overlaps
   iv Updating and network compression
   v Tracking progress in terms of time, cost and resources
   vi Cost reporting.

   The 'project management' module integrates time, resource and cost management. It can be used by single-project users as well as organisations managing multiple projects across multiple job sites. Users are able to track and analyse project performance through the various control functions.

   The P6 'project management' module centralises an organisation's resource management function by allowing each project to assign resources from the organisation's common pool of resources. The resource usage can then be effectively managed and controlled through the Web-based timesheet management application. In addition, this module also enables the user to perform integrated risk management, issue tracking and management by threshold. The tracking feature enables the user to perform cross-project cost, schedule and earned value updates regularly throughout the duration of the project, and enables customised production of various reports (Oracle Corporation 2009).

2 'Methodology management', which is a system for authoring and storing methodologies or project plan templates. 'Methodology management' allows an organisation to gather its 'best practices' and create custom project plans, refining methodology activities, estimates and other information with each new project (Oracle Corporation 2009).
3 'Timesheets module', which is a Web-based inter-project communication and timekeeping system. This module helps project participants focus on the work at hand with a simple cross-project to-do list of their upcoming assignments. Project participants can also record changes and timecards for management approval, which in turn involves project participants in the project programme and allows project managers to focus on and make the crucial project decisions (Oracle Corporation 2009).
4 'Claim Digger' is a module that provides the capability to compare two projects, or a project and its associated baseline. 'Claim Digger' helps determine what data has been changed from the schedules. Its best use is realised when an 'extension of time claim' requires assessing (Oracle Corporation 2009).

This chapter only looks at the 'project management module' of Primavera P6.

Primavera P6 gives the planner an opportunity to accommodate different work patterns for an unlimited amount of calendars. There are three calendar pools available in P6 which can be applied to resources or activities, or globally to the entire project. These are:

- The global calendar, which is a pool of calendars that apply to all projects in the database
- The project calendar pool, which is a separate pool of calendars for each project in the organisation
- The resource calendar pool, which is a separate pool of calendars for each resource.

Primavera P6 requires users develop a hierarchy of work to be accomplished in the form of an enterprise project structure (EPS), organisational breakdown structure (OBS) and work breakdown structure (WBS). The EPS is used to organise and manage the projects in an organisation. The EPS is arranged in a hierarchy and can be subdivided into as many levels or nodes as needed. The OBS is a hierarchical arrangement of a project's management structure. User access and privileges to projects within the EPS are controlled through the OBS hierarchy. The WBS is a hierarchical arrangement of the activities produced within project. The project is the highest level of the WBS while an individual activity is the lowest. The relationship of the EPS, OBS and WBS is shown in Figure 7.1.

With regard to the WBS, Primavera P6 can create an unlimited amount of project structure levels. Figure 7.2 shows three levels of WBS of a shopping centre development project with appropriate codes. Naturally, more levels of WBS can be created as they are required. By sorting activities in a schedule according to their WBS codes, the planner can review information at different levels of detail.

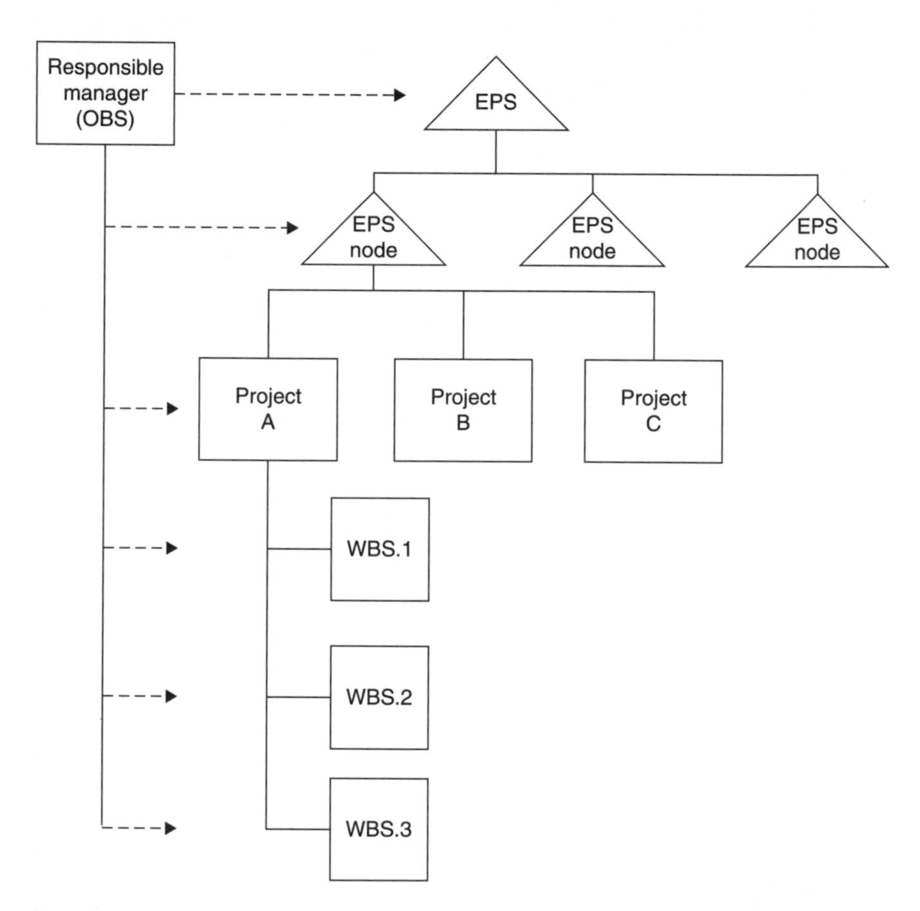

Figure 7.1 Hierarchical diagram showing EPS, OBS and WBS relationships (taken from Primavera version 6.2.1, reference manual page 27).

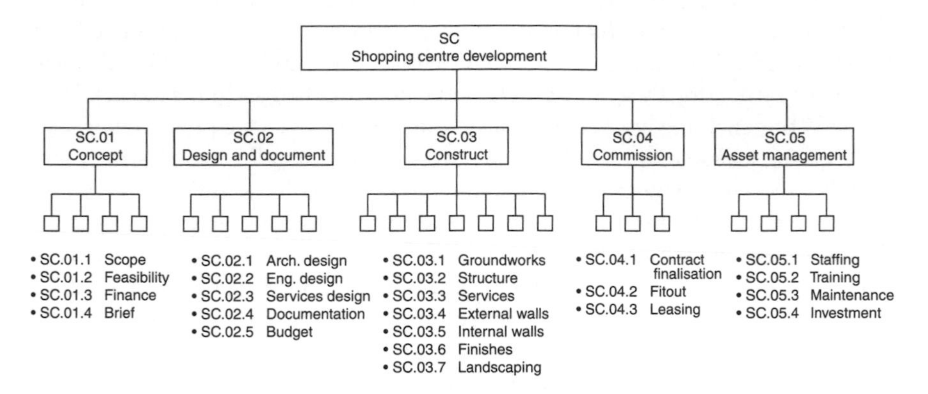

Figure 7.2 A WBS for a shopping centre development project.

When setting up the WBS, anticipated dates, budgets and spending plans can be established at a high level in the WBS to indicate when the work should occur and what the planned budgets and monthly spending should cost. Once budget amounts and funding information are set for a WBS, the information can be shared across projects. Earned value calculations can be specified for each WBS element; this is done through the use of weighted WBS milestones assigned at the required WBS level.

In addition to the EPS and WBS, code structures can also be created. Codes allow the planner to categorise projects, activities and resources with similar attributes. Once created, the planner can group, sort and filter the project based on codes.

Project codes are useful for tracking information on individual projects in the project group. By assigning project codes, the planner can classify activities into specific groups for organising, sorting and filtering information in projects. An unlimited amount of hierarchical project codes can be specified in P6.

Activity codes are assigned to individual project activities and represent broad categories of information, such as design, authority approvals, project stage, or location. Users can define specific values as part of each activity code, these specific values further describe that activity code category. For example, if a project has four separate buildings, the planner can create an activity code called 'building' with values for each individual building such as 'Main building one', 'Main building two', 'West building' and 'Car park'. The planner can then associate activities with a specific location, such as 'Bulk excavation' for each building.

Primavera P6 has a powerful cost control function that allows the planner to establish budgets, cash flows and earned value calculations necessary to complete the project. Prior to commencing a project, the planner, in consultation with the project cost estimator, would prepare separate estimates for labour, materials, plant/equipment and other resources. This information would then be input to the cost-accounting structure. As project requirements change over the course of the project, the resource and financial information can be adjusted to include for those changes. The project funding information can also be controlled through top-down estimating by applying weighted values and budget limits to each level of the WBS, enabling financiers to have total control over the project budgets.

Resource schedules are easily created in Primavera P6. P6 includes the functionality required to create a resource hierarchy that reflects the user's organisation resource structure and supports the assignment of resources to activities. There are unlimited hierarchical resource codes available for grouping resources. The project resource plan integrates resources, costs, and the schedule and allows the project manager to maintain effective control of the project. For each resource, availability limits, unit prices, and calendar information can be set. Once established, the planner can analyse the resource allocation and adjust the project plan to avoid over-allocation, peaks and valleys through resource levelling (Oracle Corporation 2009).

7.3 Scheduling a residential project using Primavera P6

The project in question is a small residential building. A list of activities including durations and resources is given in Table 7.1.

Resources can be assigned to activities either as lump-sum values or as rates per time-unit. For example, the activity 'Demolish' is assigned the supervisor's time in the form of a lump sum, that is, one person-day. Because the activity 'Demolish' is expected to take three days to complete, Primavera P6 assigns one-third of the supervisor's time to each day of the activity's duration. In comparison, the labourer is allocated to the activity 'Demolish' in the form of a rate, that is, one person per day. Primavera P6 then assigns one labourer for each day of the activity's duration, irrespective of how long the activity actually takes to complete.

The logic of the construction sequence is defined in the precedence schedule in Figure 7.3.

7.3.1 Create time schedule

Creating a new project and entering activities into Primavera P6 is a relatively simple task. Activities can be entered in a bar chart view that displays the activity form or in an activity network view.

For this simple example, neither project nor activity codes are considered. Activity IDs have been created as two-digit numerical codes. After all the activities have been logged in, they are then linked to form the required relationships. At this stage of the schedule development, only finish-to-start links are used. Overlaps will be introduced later in this chapter.

The schedule will now be calculated using the time scheduler, which is only concerned with tasks while ignoring committed resources. Primavera P6 has

Table 7.1 The project data

Activities	*Project resources*						
	Time (days)	*Labourer*	*Carpenter*	*Subcontractor*	*Supervisor*	*Plant*	*All labour*
5 Demolish	3	1/Day		5/Day	1 Person-day	1 Bulldozer/day 1 Truck/day	7
10 Concrete	6	3/Day	1/Day		4 Person-days	2 Backhoes/day 1 Pump/day 1 Vibrator/day	5
15 External walls	6			7/Day	4 Person-days	1 Mixer/day 3 Barrows/day 1 Scaffold/day	8
20 Roof	4	1/Day	2/Day	3/Day	2 Person-days		7
25 Timber floor	3	2/Day	4/Day		2 Person-days		7
30 Internal walls	3	2/Day	4/Day		2 Person-days		7
35 Plumbing	7			14 Person-days	2 Person-days		3
40 Electrical	6			12 Person-days	2 Person-days		3
45 Floor finish	2	2/Day	3/Day		1 Person-day		6
50 Ceiling	3			2/Day			2
55 Finishing	7	2/Day	1/Day	3/Day	5 Person-days		7

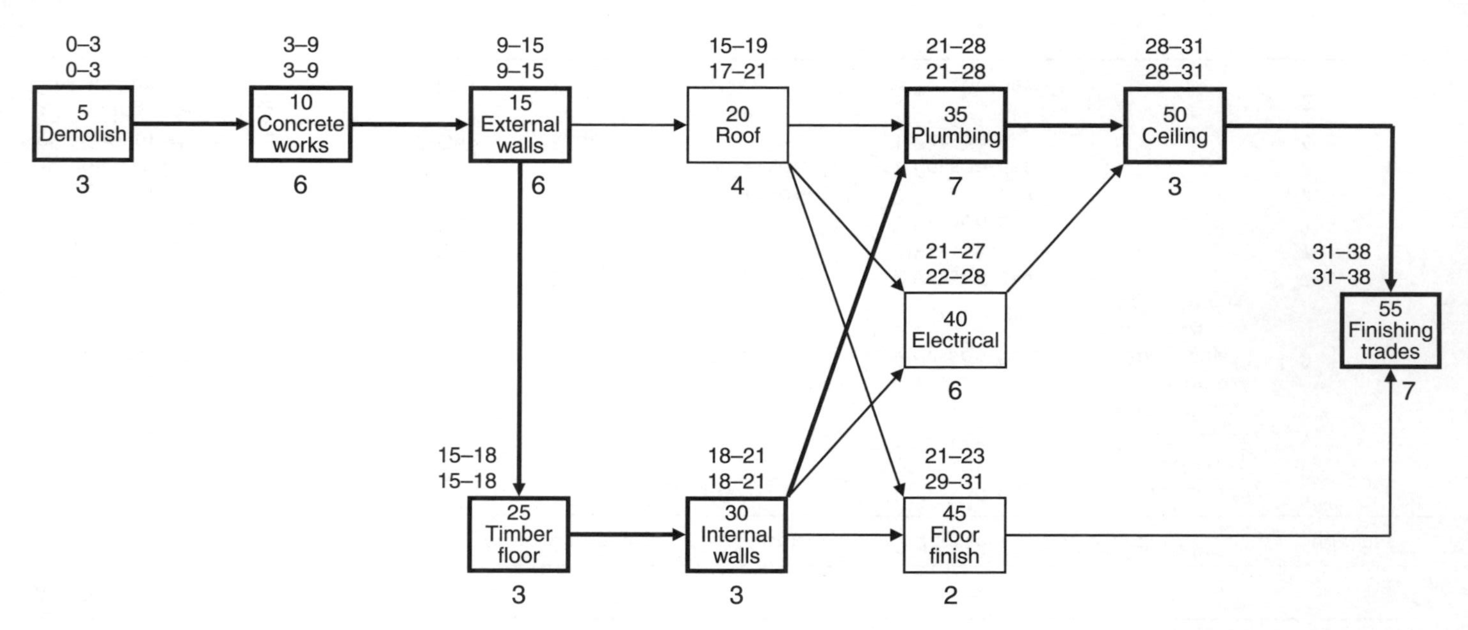

Figure 7.3 A manually produced and calculated precedence schedule of the residential project.

calculated the project to be completed on day 39. Since Primavera P6 starts the project on day 1, week 1, the actual duration of the project in days is 38. The resulting time schedule can be viewed/printed in a number of different formats such as a linked bar chart, a time-logic diagram, activity network or a table. Figure 7.4 displays the time schedule in the form of a linked bar chart.

7.3.2 Create resource schedule

Primavera P6 levels resources according to the following rules:

- Initially, resources are levelled within their normal limits by using up available total float but maintaining the project duration
- If resources cannot be levelled within their normal limits, Primavera levels them within their total limits while maintaining the project duration
- If Primavera P6 cannot level resources within their total limits, it will then extend the project duration.

A 'total limit' of a resource is the maximum volume of that resource available for the execution of the work.

Limits imposed on the volume of required resources are specified in Table 7.2.

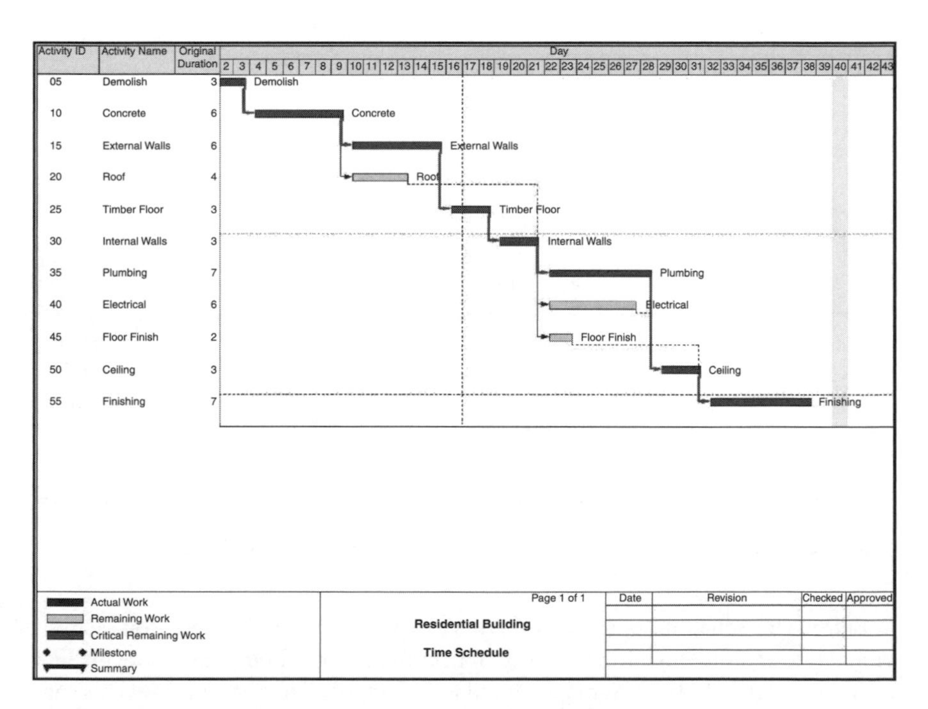

Figure 7.4 The time schedule in the form of a linked bar chart.

Table 7.2 Normal and total resource limits

Resources	*Normal limit*	*Total limit*
All labour	10	15
Backhoe	2	2
Barrow	3	3
Bulldozer	1	1
Carpenter	4	5
Labourer	3	4
Mixer	1	1
Pump	1	1
Scaffold	1	1
Subcontract labour	8	10
Supervisor	2	3
Truck	1	1
Vibrator	1	1

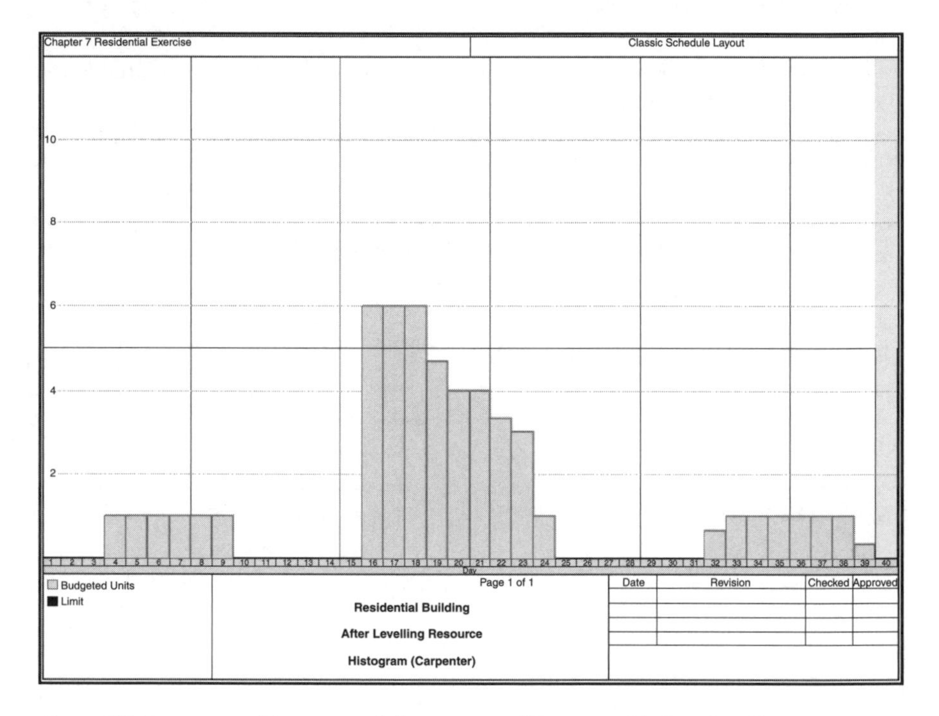

Figure 7.5 A resource histogram of the resource 'Carpenter' before levelling.

Before the schedule is resource-levelled, it is worth browsing through individual resources to examine their distribution and note their demand levels. For example, a resource histogram generated for 'Carpenter' in Figure 7.5 reveals that the demand for carpenters between weeks 3 and 4 exceeds both the normal and total limits. The activities causing this excessive demand are 'Timber floor' and 'Roof'. Unless more carpenters are assigned to one or both of these activities, Primavera P6 will delay the start of the non-critical activity 'Roof' until the total resource limit is met.

Let's resource-level the schedule based on the defined limits of resources using the resource leveller in Primavera P6. The calculated project completion date is now day 43. Examination of the resource schedule in Figure 7.6 reveals that by observing the resource limits of the resource 'Carpenter', Primavera P6 delayed not just the activity 'Roof' but also the critical activity 'Internal walls'. Consequently, all the other activities scheduled beyond the activity 'Internal walls' were delayed.

Figure 7.6 also includes baseline schedules, which are a complete copy of the original project schedule. A baseline schedule prior to the resource levelling exercise is shown in Figure 7.4. The 'baseline' is used as a comparison with the current schedule. Before the planner updates a schedule for the first time, a baseline plan is typically produced. 'Baseline' provides a target against which the planner is able track a project's cost, schedule and performance.

The resource histogram of the resource 'Carpenter' after resource levelling is given in Figure 7.7. It shows that the demand for the resource is now well within its normal limit.

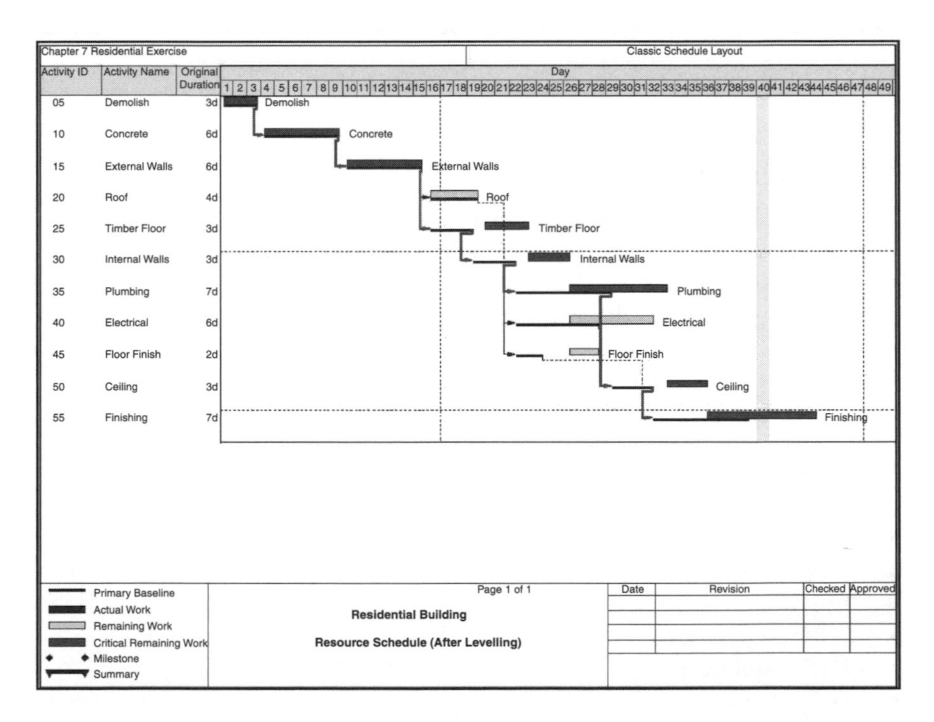

Figure 7.6 The resource schedule in the form of a linked bar chart with a baseline.

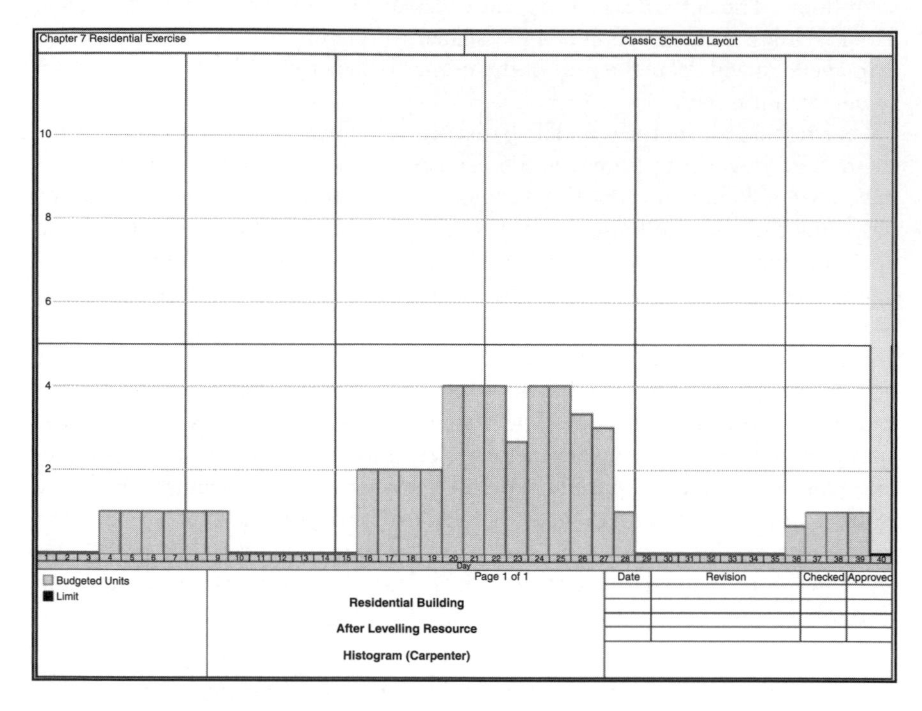

Figure 7.7 A resource histogram of 'Carpenter' after levelling.

7.3.3 Cost estimating

For this residential project, five account categories have been set up as follows:

- L – labour
- C – contractor and subcontractors
- S – supervisor
- P – plant and equipment
- M – materials.

Cost accounts have been set up for the following resources:

- 1201 – labourer
- 1202 – carpenter
- 1203 – subcontractor
- 1204 – supervisor
- 1205 – bulldozer
- 1206 – truck
- 1207 – backhoe

- 1208 – concrete pump
- 1209 – vibrator
- 1210 – mixer
- 1211 – wheelbarrow
- 1212 – scaffolding
- 1213 – materials.

Individual cost components have been estimated for each activity in the project. They are given in Table 7.3.

The cost rates for labour and plant are given in Table 7.4.

Primavera P6 provides a wide range of cost reports including cost control, earned value, tabular and cost loading reports. (An example of a cost report detailed by activity is given later, in Table 7.6.)

7.3.4 Updating the schedule

As the project gets under way, the project manager regularly monitors performance and compares it with the planned or budgeted performance. From deviations and trends that have been detected and analysed, the project manager updates the schedule accordingly.

The process of schedule updating is a relatively simple task in Primavera P6. The revised data on duration of activities, resource requirements and costs, related to a particular point in time, are logged in. Let's assume that the schedule is being reviewed on day 27. The project manager's assessment of progress achieved to date, and the estimates of future progress in terms of time and cost, are given in Table 7.5.

Once the revised time and cost information is logged in, the schedule is recalculated and new reports generated. A linked bar chart in Figure 7.8 and a cost report sample in Table 7.6 provide information on the outcome of the review on day 27.

7.4 Overlapping models in Primavera P6

In addition to the default FTS link, Primavera P6 is able to model STS, FTF and STF links, as well as a compound STS and FTF link. These overlapping models were discussed in detail in Chapter 5.

CPM schedules of small construction projects commonly involve FTS links only. Overlaps are rare due to their small size and a relatively small number of activities. While not entirely necessary, overlaps have been introduced into the residential project in question for demonstration purposes. They are illustrated in Figure 7.9. The manual calculation of the time schedule shows its duration to be 27 days.

Primavera P6 adds overlaps by selecting an appropriate overlap from the 'Activity Detail' menu. A compound STS and FTF overlap requires the creation of two links, one for the STS and the other for the FTF link.

Table 7.3 The project cost data

Activities	*Costs*						
	Labourer	*Carpenter*	*Subcontractor*	*Supervisor*	*Plant*	*Materials*	*Total cost*
5 Demolish	$420		$3,000	$160	$1,500	$320	$5,400
10 Concrete	$2,520	$900		$640	$530	$1,300	$5,890
15 External walls			$8,400	$640	$1,080	$6,750	$16,870
20 Roof	$560	$1,200	$2,400	$320		$4,150	$8,630
25 Timber floor	$840	$1,800		$320		$1,890	$4,850
30 Internal walls	$840	$1,800		$320		$5,240	$8,200
35 Plumbing			$2,800	$320		$1,740	$4,860
40 Electrical			$2,400	$320		$1,520	$4,240
45 Floor finish	$560	$900		$160		$2,670	$4,290
50 Ceiling			$1,200			$2,050	$3,250
55 Finishing	$1,960	$1,050	$4,200	$800		$3,250	$11,260
Total	**$7,700**	**$7,650**	**$24,400**	**$4,000**	**$3,110**	**$30,880**	**$77,740**

Table 7.4 The cost rates for labour and plant

Resources	*Cost per day*
Backhoe	$150
Barrow	$10
Bulldozer	$300
Carpenter	$150
Labourer	$140
Mixer	$50
Pump	$200
Scaffold	$100
Subcontract labour	$200
Supervisor	$160
Truck	$200
Vibrator	$30

Table 7.5 Data on progress of the project on day 27

Activities fully completed

Activity number	*Actual duration*	*Actual cost ($)*
5	3	5,400
10	6	6,200
15	6	16,970
20	4	9,140
25	3	5,060
30	3	8,500

Activities in progress

Activity number	*Time to complete*	*Cost to date*	*Cost to complete*
35	6	800	4,170
45	1	2,650	1,860

Activities not yet started

Activity number	*Revised duration*	*Revised cost ($)*
40	5	4,240
50	3	3,250
55	7	11,260

For the residential project in question, overlaps were added to the resource schedule. When the schedule was recalculated, its duration remained unchanged (day 42) (see Figure 7.10). This is because the resource schedule is calculated within the limits of available resources. In comparison, the overlapped time schedule in Figure 7.9 had its duration reduced to 27 days.

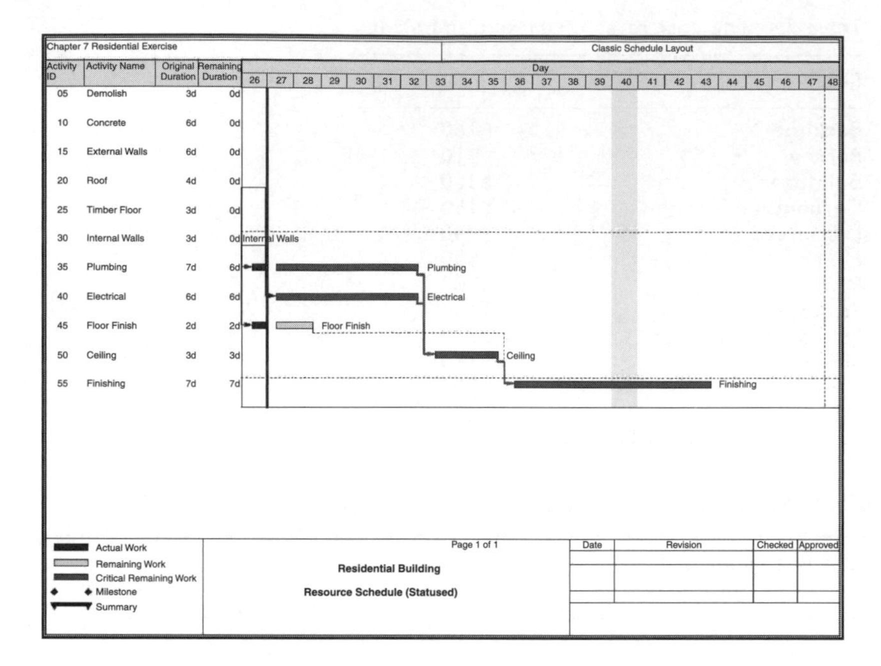

Figure 7.8 The project schedule in the form of a linked bar chart on day 27.

7.5 BIM software

The proffered communication medium for project information has traditionally been two-dimensional (2D) drawings. With the introduction of three-dimensional (3D) architectural design software such as Revit and Bentley, the use of 3D graphical models between planning and design phases has rapidly increased in popularity. The introduction of 3D design documentation has initiated the growth of five-dimensional (5D) models which integrate a 3D drawing with time and cost estimate parameters. The benefits of 5D models are yet to be fully realised in the international construction industry due to the relatively new introduction of 5D technology (Goedert and Meadati 2008).

The method of integrating the construction schedule with 3D CAD file requires:

- The graphical 3D CAD file to be created in the architectural design software
- The construction schedule to be created in Microsoft Project, Primavera or Asta Powerproject
- The construction schedule converted to a file type that the design software can read.

Table 7.6 A sample of a cost report for the project review on day 27

Cost Control Report

Activity Name

Resource ID	Name	Cost Account	% Complete	Budgeted Cost	Actual Cost	At Completion Cost
0.5	0.5 Demolish	RD 0d				
	Labourer.Labourer	1201	100%	$420	$420	$420
	Bulldozer.Bulldozer	1205	100%	$900	$750	$750
	Subcontract Labour.Subcontract Labour	1203	100%	$3,000	$3,000	$3,000
	Supervisor.Supervisor	1204	100%	$160	$160	$160
	Truck.Truck	1206	100%	$600	$750	$750
	Materials.Materials	1213	100%	$320	$320	$320
Subtotal			0%	$5,400	$5,400	$5,400
10	10 Concrete	RD 0d				
	Labourer.Labourer	1201	100%	$2,520	$2,520	$2,700
	Carpenter.Carpenter	1202	100%	$900	$900	$950
	Supervisor.Supervisor	1204	100%	$640	$640	$640
	Pump.Pump	1208	100%	$200	$200	$200
	Backhoe.Backhoe	1207	100%	$300	$300	$330
	Materials.Materials	1213	100%	$1,300	$1,300	$1,350
	Vibrator.Vibrator	1209	100%	$30	$30	$30
Subtotal			0%	$5,880	5,880	$6,200
15	15 External Walls	RD 0d				
	Supervisor Supervisor	1204	100%	$640	$640	$640
	Subcontract Labour.Subcontract Labour	1203	100%	$8,400	$8,400	$8,400
	Mixer.Mixer	1210	100%	$300	$300	$300
	Barrow.Barrow	1211	100%	$180	$180	$180
	Scaffold.Scaffold	1212	100%	$600	$600	$660
	Materials.Materials	1213	100%	$6,750	$6,750	$6,890
Subtotal			0%	$18,870	$18,870	$17,070
25	25 Timber Floor	RD 0d				
	Labourer.Labourer	1201	100%	$840	$840	$860
	Carpenter.Carpenter	1202	100%	$1,800	$1,800	$2,100
	Supervisor.Supervisor	1204	100%	$320	$320	$320
	Materials.Materials	1213	100%	$1,890	$1,890	$1,890
Subtotal			0%	$4,850	$4,850	$5,170
30	30 Internal Walls	RD 0d				
	Labourer.Labourer	1201	100%	$840	$840	$840
	Carpenter.Carpenter	1202	100%	$1,800	$1,800	$1,850
	Supervisor.Supervisor	1204	100%	$320	$320	$320
	Materials.Materials	1213	100%	$5,240	$5,240	$5,530
Subtotal			0%	$8,200	$8,200	$8,540
35	35 Plumbing	RD 6d				
	Subcontract Labour.Subcontract Labour	1203	14.29%	$2,800	$400	$2,800
	Supervisor.Supervisor	1204	14.29%	$320	$46	$320
	Materials.Materials	1213	14.29%	$1,740	$249	$1,850
Subtotal			0%	$4,880	$886	$4,870
40	40 Electrical	RD 6d				
	Subcontract Labour.Subcontract Labour	1203	0%	$2,400	$0	$2,400
	Supervisor.Supervisor	1204	0%	$320	$0	$320
	Materials.Materials	1213	0%	$1,520	$0	$1,520
Subtotal			0%	$4,240	$0	$4,240
45	45 Floor Finish	RD 1d				
	Labourer.Labourer	1201	0%	$560	$560	$600
	Carpenter.Carpenter	1202	0%	$900	$450	$1,050
	Supervisor.Supervisor	1204	0%	$160	$80	$100
	Materials.Materials	1213	0%	$2,670	$1,335	$2,760
Subtotal			0%	$4,290	$2,425	$4,610
50	50 Ceiling	RD 3d				
	Subcontract Labour.Subcontract Labour	1203	0%	$1,200	$0	$1,200
	Materials.Materials	1213	0%	$2,050	$0	$2,050
Subtotal			0%	$3,250	$0	$3,250

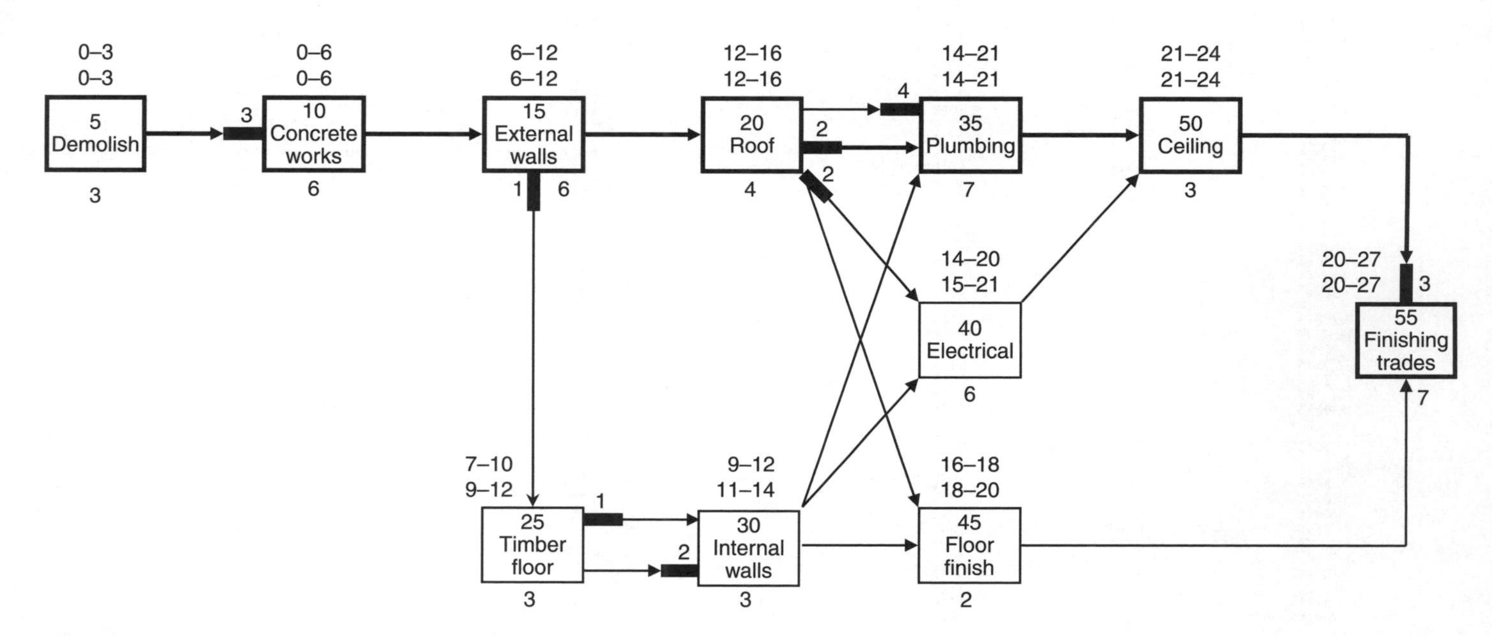

Figure 7.9 An overlapped schedule of the project.

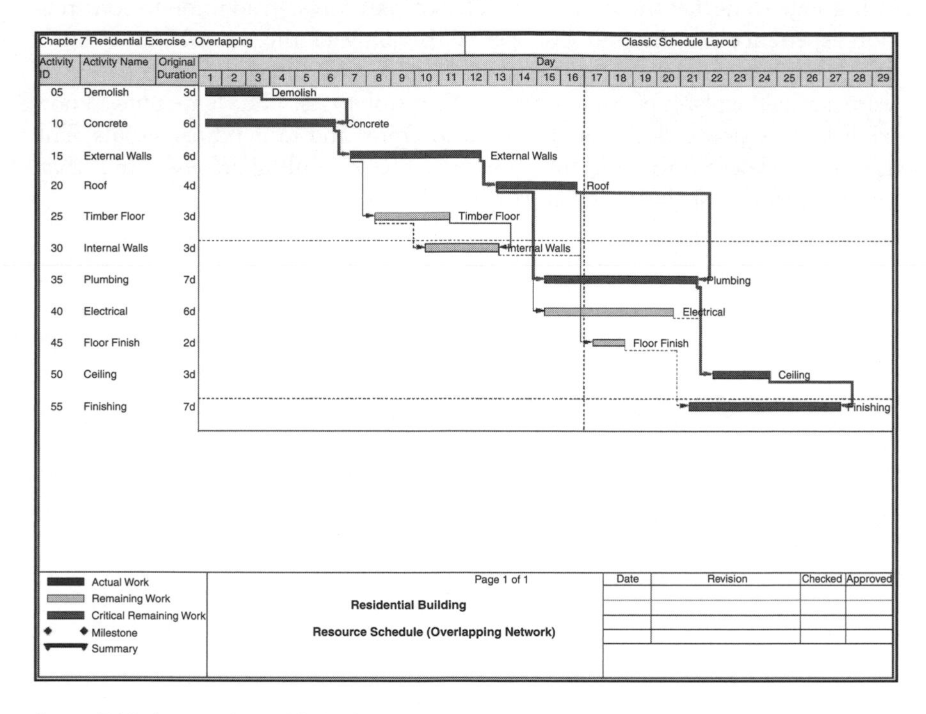

Figure 7.10 An overlapped linked bar chart.

Each construction schedule activity is then manually linked to a 3D graphic model objects within the architectural design software. The automated 5D model can then be created using the architectural design software.

As the level of information required to create the 5D model is quite detailed, it is often difficult to create a 5D automated model in the concept and design phases of a project, therefore limiting the effective use of 5D models. Given that 5D modelling is in early phases of its development, it is expected that its application in the construction industry will increase when the user interface and ease of use is improved by software developers.

7.6 Summary

This chapter examined some aspects of critical path scheduling using Primavera P6 software with regard to a small residential project. It explained the preparation and calculation of time and resource schedules, it defined activity and cost codes and it generated samples of cost reports. It also demonstrated the process of schedule updating. The chapter also illustrated the use of overlapping models in critical path scheduling and introduced the idea of 5D modelling.

It is important that those who use CPM software have, in addition to construction experience and the ability to operate the software, theoretical knowledge of critical path scheduling. This is essential for effective interaction with the software during the planning, monitoring and control stages. The planner must know what information to log in and how to interpret and respond to outputs. The planner is expected to be in full control of the scheduling process rather than being controlled or constrained by the CPM software.

Chapter 8

Critical chain scheduling

8.1 Introduction

The concept of critical path was discussed in the preceding chapters. It is the most commonly used scheduling method in the construction industry. In recent years, however, the proponents of the 'theory of constraints' (TOC) have questioned the appropriateness of the critical path method (CPM) in project management, arguing in favour of an alternative approach derived from TOC, which is known as 'critical chain scheduling' (CCS). This approach may also be referred to as 'critical chain management' (CCM). In this chapter, the former term is used.

The purpose of this brief chapter is to examine the concept of CCS, highlight its differences from the CPM, and describe its benefits.

8.2 Shortcomings of the critical path method

The proponents of CCS, including Goldratt (1997) and Newbold (1998), have identified a number of shortcomings of CPM that CCS is able to overcome. These are mainly related to excessive use of contingencies in time estimation, task orientation, frequent shifting of the critical path, and multi-tasking.

In traditional CPM scheduling, planners derive duration of activities deterministically as an average time estimate that has a 50 per cent probability of being achieved. This 'risky' estimate is then made safe by the inclusion of an allowance or contingency for uncertainty, which the planner adds to each activity. This contingency represents the planner's subjective perception of the risk to which each activity is exposed. Over hundreds of activities in a schedule, these individual activity contingencies add up to a substantial time contingency. Thereafter, top management of the firm usually adds another contingency to the schedule, which is intended to cover uncertainties surrounding the entire project. This contingency is referred to as project contingency and is commonly expressed as a percentage of the project period.

It would seem that with two layers of contingency built into a schedule, the probability of delays in delivering projects on time should be small. However, the reality is different, and construction projects continue to overrun on time. The proponents of CCS argue that excessive contingencies create a feeling of 'safety' among the project team members and this may lead to complacence.

The proponents of CCS point to task orientation as another weakness of CPM. Indeed, CPM is task- or activity-oriented. The emphasis is placed on completing individual activities as planned. The project manager hopes that by accomplishing individual activities as scheduled, the project will be completed on time. Having sufficient time contingencies and float in non-critical activities may, however, lead to complacence; people's attention tends to shift away from those activities that have 'time safety' built into them to more urgent tasks. They may even reallocate resources away from such safe activities to those that need them urgently. This is known as 'student syndrome', which refers to the common practice among students of not attending to assignments until a last few days before their submission. The lack of focus on non-urgent tasks results in the loss of float. Non-critical activities then become critical and their further delay causes the project to overrun. It may be argued that the focus on tasks diminishes the focus on the overall completion date of the project, which is the most important issue as far as the client is concerned. A statement made by Patrick (1999: 5) reinforces this view: 'We protect our project due dates by protecting task due dates with safety. Then, from the point of view of the project, we waste that safety due to the comfort it provides, and put the project promise in jeopardy'.

Efficient use of committed resources is an important aspect of critical path scheduling. This is achieved through resource levelling. When a schedule has been resource-levelled, each activity in the schedule becomes critical in terms of committed resources. The loss of float in one specific activity may not by itself cause a delay to the overall schedule but it may alter the demand level for a resource that has been allocated to it. This may lead to extra costs, multi-tasking of committed resources, and/or a delay to the schedule. If resource allocation to critical activities is insufficient, such activities will be delayed unless more resources are urgently provided. This again will increase the cost and cause multi-tasking since resources from other activities will be diverted to activities in distress.

A resource-levelled schedule contains both task and resource dependencies. However, only task dependencies are shown in a critical path schedule as links. Resource dependencies are formed by resource levelling. Omission of some dependencies is likely to result in unexpected delays in activities. For example, a formwork subcontractor who is designing slipform equipment for the construction of the lift shafts realises that information on the location of reinforcing bars in concrete walls has not been provided. This is because a dependency between structural design and slipform design was omitted. The outcome of this omission is that slipform design will most likely be delayed if the structural engineer's resources cannot be mobilised immediately to address this problem. The most likely outcome is that the structural engineer will divert some of the design resources from other tasks to solving this problem. This potentially leads to multi-tasking of resources, which increases the risk of further delays.

References have previously been made to multi-tasking as a potential cause of delays. Multi-tasking is a common feature of CPM scheduling. It is characterised by employing resources across a number of different activities in a given period in order to achieve their maximum use. However, without clear priority of employment of such resources, as a safeguard, durations of activities are commonly made longer to ensure that the resources are able to accomplish the work. The extra time added to activities to compensate for multi-tasking may be lost due to the task orientation of the CPM, which was briefly discussed earlier. A multi-tasked resource such an electrician, who is responsible for a wide range of activities associated with the installation and supply of the electrical service, is likely to be directed first to urgent tasks. If delays occur in the execution of such urgent tasks, non-urgent tasks lose their float and also become urgent. It follows that multi-tasking is likely to increase a risk of delays.

Knowledge of the location of the critical path is the essence of critical path scheduling. The project manager concentrates management efforts on preventing delays in critical activities in order to maintain the project schedule. When float is lost, non-critical activities become critical and form an additional critical path. Any further delay in their duration results in the loss of the original critical path. By the time the project manager contains the problem, which commonly requires reallocation of resources or injection of additional resources, another shift in the critical path may occur. The project manager's work then involves 'putting out fires' rather than managing. The outcome is rising cost and delay in completing the project.

By focusing largely on critical activities, the project manager may not recognise a high risk of future delays associated with non-critical paths. Just because non-critical activities have float does not guarantee that the amount of float is enough to safeguard against future delays.

The proponents of CCS believe that project schedules can be substantially reduced or the risk of time overruns minimised by the application of CCS. Before explaining how CCS works, let's first define the TOC from which CCS has emerged.

8.3 Theory of constraints

TOC was developed by Dr Eli Goldratt of the Avraham Y. Goldratt Institute (Goldratt 1997). It is a philosophy or a thinking process that helps management to develop appropriate solutions to problems that in turn improve the performance of projects. TOC is an integrated problem-solving methodology based on a holistic approach to problem solving. According to the TOC approach, components of the system being managed must be aligned and integrated with higher-level systems of which they are components. This requires effective management of project tasks and resources in consideration of defined project goals.

TOC is based on the premise that every organisation or system has at least one weak link or constraint that inhibits its ability to meet its goals. This weak link must be found and strategies developed for either eliminating it or learning to manage the work around it. The constraint is thus removed and the project is then uninhibited in achieving its goals.

The TOC thinking process provides specific tools for identifying such constraints and developing appropriate solutions. Since it brings many people together, it is also an effective communication tool that promotes team building. The TOC thinking process approaches problem solving strategically and tactically. Strategic problem-solving attempts to find breakthrough solutions and requires a paradigm shift that may change the way in which the organisation operates. Tactical problem-solving provides tools for resolving day-to-day problems; these tools are mainly communication, conflict resolution, empowerment and team-building tools.

CCS is the TOC-based technique for managing projects. Unlike the traditional CPM, CCS adopts a holistic approach to developing a project schedule. The main features of CCS will now be briefly discussed.

8.4 Critical chain scheduling

The aim of CCS is to improve project performance by achieving a scheduled completion date, which in turn would very likely result in improved cost performance. Since it is a TOC-based technique, it is applied in a systematic and co-ordinated manner so that constraints preventing the achievement of project goals are identified and eliminated. This approach will also ensure that all the important dependencies among project activities have been identified and correctly built into a schedule.

From the CCS viewpoint, task orientation of the CPM is a constraint. Another constraint is the excessive use of time contingencies that are added to individual activities as well as to the entire project. Proponents of CCS argue that these time contingencies and float may result in complacence, which may then turn into delays, multi-tasking and extra costs. CCS attempts to shift focus away from achieving individual tasks to meeting the scheduled completion date. It uses a radical two-prong approach:

- It takes the focus away from due dates of activities since most of them will not be completed on the scheduled date anyway. It adopts an approach that allows the work associated with tasks to take as long as it takes.
- It removes time contingencies from individual activities.

How can a target date be met without exercising strict control over due dates of individual tasks and without any time contingency? Activities on a critical path and associated resources allocated to such activities must be given close attention since they control the project's duration. Let's refer to such critical activities as critical chain tasks. It is important to ensure that critical chain resources are available to start work when the immediately preceding activity has been finished. Rather than specifying the completion date of such a preceding activity, which without a time contingency has at most only 50 per cent probability of being met, it may be possible to arrange for a succeeding resource to be on standby and start working after the preceding activity has actually been completed. In this case the project manager would negotiate the amount of advance warning that the succeeding resource would require before starting the work.

With little emphasis placed on task dates and no time contingencies present in individual activities, the schedule is obviously very tight. Such a schedule offers no protection against risks to which the project may be exposed. But rather than safeguarding the due dates of individual critical tasks, CCS safeguards the end date of the project by adding a project buffer to the chain of critical tasks. The magnitude of this buffer is expected to be much smaller than the sum of individual time contingencies that the planner adds to critical activities in critical path schedules. Patrick (1999: 8) speculates that 'we can usually cut the total protection at least in half and still be safe'.

So far the discussion has evolved around critical activities or critical chains of tasks. Non-critical chains could be managed in much the same way as critical ones, but it may be rather tedious to arrange advance alerts for resources allocated to non-critical tasks. Instead, CCS uses buffers that are added to non-critical chains. They are referred to as 'feeding buffers' because they protect the start of the critical chain task from delays caused by the feeding of non-critical chains. The critical chain is thus protected not just by the project buffer but also by the feeding buffers. The feeding buffers also safeguard against frequent shifting of the critical path.

Multi-tasking has also been identified as a constraint of CPM. CCS reduces the occurrence of multi-tasking by shortening durations of activities. This requires the work to be done by the resource without any delay. When multi-tasking is minimised, the project manager is in a better position to identify those resources that actually constrain the project.

A project schedule developed using the principles of CCS is expected to be 20–30 per cent shorter in duration than a schedule produced by the traditional CPM (Anon 2002: 2). Once the project is under way, its progress requires regular monitoring and control. Project control in CCS is known as buffer management. At regular intervals the project manager tracks the consumption or the replenishment of project and feeding buffers. The buffers are usually broken up into three segments. The first contains a time allowance that is expendable and no action is required if some or all of it is actually consumed. When a time allowance in the second segment is beginning to be consumed, the project manager is required to develop a plan of action for recovering some or all of the time lost. If part of the third segment has been expended, the project manager will implement the planned action to recover the lost time.

From this examination of key features of CCS, some important benefits of CCS emerge. These include the following (Patrick 1999):

- Shorter activity durations
- Focus on the end date of the project rather than on due dates of activities
- The use of advance warning for commencement of resources
- Prioritisation of resource use
- Minimisation of multi-tasking
- Project control by buffer management.

8.5 Comparison of critical path method (CPM) and critical chain scheduling (CCS)

The two main characteristics that differ between CPM and CCS are the method of estimating task durations and risk control techniques. When estimating task durations, CPM schedules use 'realistic task estimates' (see Chapter 11 for details of activity duration estimates) whereas CCS uses 'aggressive task estimates'. The risk control mechanism of a CPM schedule is the delay contingency applied differently by different planners (see Chapter 12 for details of risk control in CPM schedules). In CC schedules, the risk control mechanism is holistically incorporated into the schedule through the activity duration estimate process and the use of buffers. By comparing the delay contingency mechanisms of CPM and CCS, a critical comparison can be made between the two techniques, which will be measured by their use of time contingencies and float.

8.5.1 A sample case study

The objective of the case study is to compare the performance of the time contingency techniques in controlling risk of both the critical path and critical chain schedules.

8.5.2 Method

Models of both CCS and CPM will be produced based on existing literature and will be critically examined to determine how each scheduling technique deals with risk. Through the step-by-step examination of each schedule, common aspects such as task estimates, time contingency allowance in task estimates, time contingency allowance in the entire schedule, the amount of free and total float in each schedule, and the ease of producing each schedule, will be critically examined and compared.

By using the common aspects of time contingencies and float to compare each schedule, the effects that each scheduling technique has on the duration of the schedule can be assessed. The use of the areas of float and time contingencies are important as a basis of assessment, as the consumption of float and time contingencies almost always affects the final date of the project. The data used for each of the CPM and CC schedules is shown in Table 8.1.

8.5.3 Assumptions

The assumptions made for this case study are:

- Resources are unlimited in all schedules
- It will be assumed that the more accurate the estimated schedule duration is the better
- It will be assumed that uncertainty is the primary risk associated with construction schedules and that the following factors usually make up the uncertainty risk contingency: inclement weather, industry-wide disputes, latent conditions, acts and regulations, statutory bodies and suppliers of materials
- It will be assumed that the planner of the CPM schedule is neutral to risk
- All figures will be rounded up to the nearest whole number.

8.5.4 Results for the critical path schedule

The calculated CPM schedule is shown in the form of a Gantt chart at Figure 8.1. Table 8.2 outlines the total length of the CPM and hence the total duration of

Table 8.1 Case study data

Activity	*Optimistic duration*	*Most-likely duration*	*Safety factor*	*Successors*
Start	0	0	0	A, C, D, F
A	5	7	2	B
B	2	2	0	G
C	5	8	3	G
D	6	8	2	E, H, I, J
E	4	4	0	G
F	5	8	3	K
G	2	2	0	L
H	3	3	0	L
I	5	9	4	N
J	2	3	1	M
K	5	5	0	M
L	2	3	1	N
M	6	8	2	N
N	2	2	0	Finish
Finish	0	0	0	

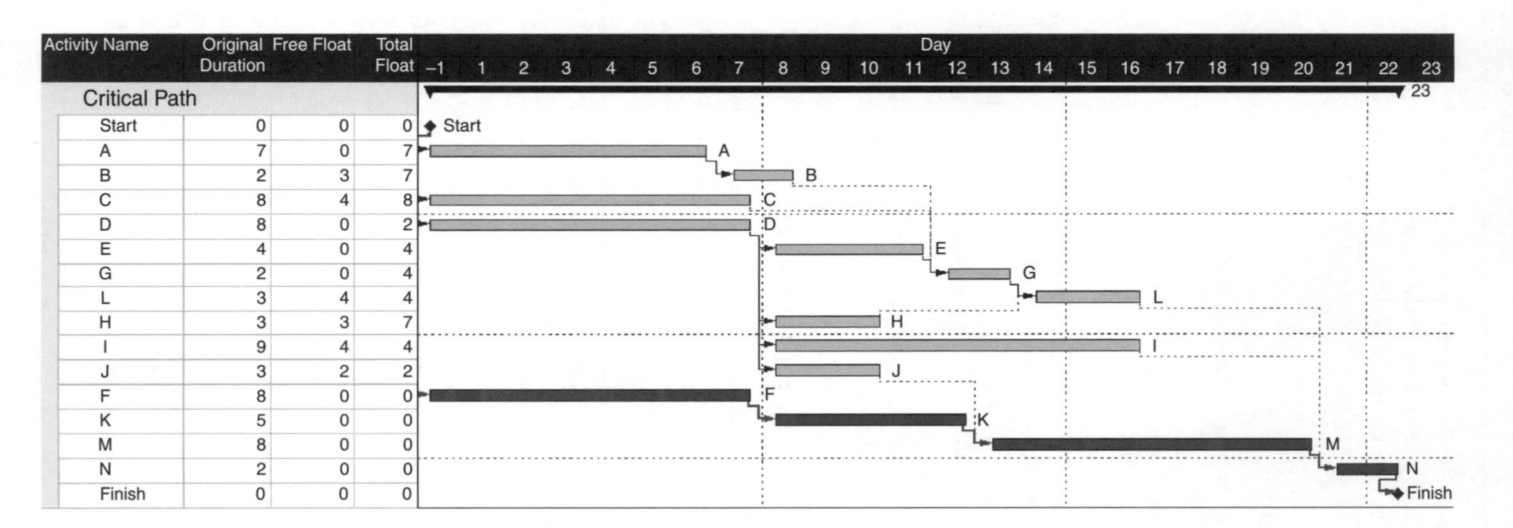

Figure 8.1 CPM schedule.

Table 8.2 Contingency in the CPM schedule

CPM number 1	*Start* →	*F* →	*K* →	*M* →	*N* →	*Finish*
Duration	0	8	5	8	2	0
Time contingency	0	3	0	2	0	0

the project. When the amount of time contingency in the CPM is divided by total length of the total duration of the project, the percentage of total time contingency on the critical path is shown.

Total schedule duration: **23 days**

Total time contingency in the CPM: 0 + 3 + 0 + 2 + 0 + 0 = **5 days**

Percentage of safety time in the CPM: 5/23 × 100% = **22%**

Table 8.3 outlines the float (free and total) in non-critical activities and the amount of safety time in non-critical activities.

Table 8.3 Float in the CPM schedule

Description	*Original duration*	*Safety time*	*Free float*	*Total float*
A	7	2	0	7
B	2	0	3	7
C	8	3	4	8
D	8	2	0	2
E	4	0	0	4
G	2	0	0	4
H	3	0	3	7
I	9	4	4	4
J	3	1	2	2
L	3	1	4	4

8.5.5 Results for the critical chain schedule

Goldratt (1997) defines the following method for developing a CC schedule:

1. The bottleneck is identified
2. Each activity is estimated with approximately 50% of the original inflated estimates
3. The schedule is programmed to start as soon as possible
4. A single block of safety time is allowed at the end (project buffer), which is equal to 50% × total safety time

5 The non-critical activities are estimated in the same way with an anticipated safety bucket of time at the end of each path feeding into the CC, to make sure that the CPM is not affected (feeding buffer).

Step 1 – the bottleneck

The bottleneck in the CCS will be the longest chain once the schedule is calculated using the optimistic activity duration estimates; see Figure 8.2.

Step 2 – activity estimates

The main assumption used in the CCS is to use 'aggressive activity estimates', i.e. a 50 per cent confidence level of finishing each activity in the noted time. 'Realistic activity estimates' usually have a 90 per cent confidence level of finishing each activity in the noted time, which includes the risk contingency. The aggressive activity estimates are those shown as 'optimistic' in Table 8.4.

Step 3 – initial CC schedule

Figure 8.2 shows all activities starting as soon as possible with 'optimistic' task durations. No allowance for risk contingency has been made. The project and 'feeding buffers' now need to be built into the schedule.

Step 4 – project buffer

In a CC schedule the planner should leave a single block of safety time, a 'project buffer', at the end, which is equal to 50% × total longest chain safety time (safety time is equal to realistic duration – aggressive duration). Therefore:

Project buffer = (Start + F + K + M + N) × 50% = **X**

Project buffer = (0 + 3 + 0 + 2 + 0) × 50% = **2.5**

Project buffer = **2.5**

Table 8.5 shows the calculation logic of the project buffer and Figure 8.3 displays the CC schedule with the project buffer inserted between the terminal activity and the dummy finish.

Step 5 – feeding buffer

The non-critical activities are estimated in the same way as the project buffer; therefore an anticipated safety block of time is placed at the end of each path feeding into the CC schedule to make sure that the CPM is not adversely affected (feeding buffer).

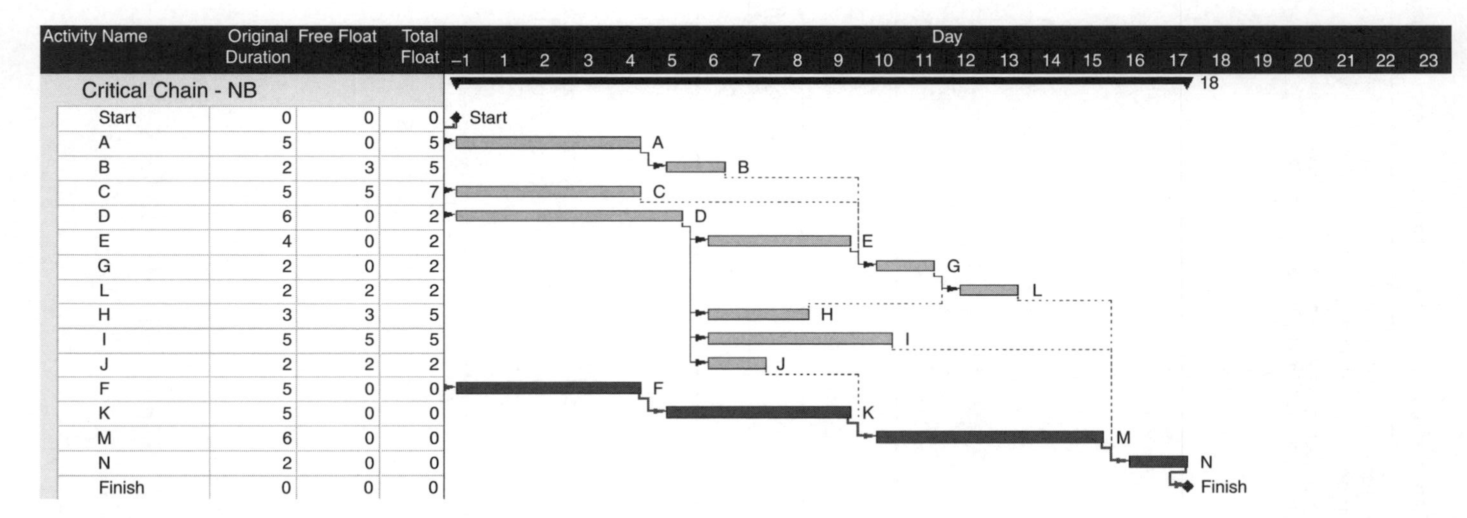

Figure 8.2 Schedule calculated with aggressive/optimistic activity durations.

Table 8.4 Critical chain activity calculation with critical chain activities highlighted

Activity	*Optimistic duration*	*Most-likely duration*	*Safety factor*	*Successors*
Start	0	0	0	A, C, D, F
A	5	7	2	B
B	2	2	0	G
C	5	8	3	G
D	6	8	2	E, H, I, J
E	4	4	0	G
F	5	8	3	K
G	2	2	0	L
H	3	3	0	L
I	5	9	4	N
J	2	3	1	M
K	5	5	0	M
L	2	3	1	N
M	6	8	2	N
N	2	2	0	Finish
Finish	0	0	0	

Table 8.5 Project buffer calculation

Critical chain	*Start* →	*F* →	*K* →	*M* →	*N* →	*Project buffer* →	*N*
Safety time	0	3	0	2	0	**2.5**	N/R

N/R: not relevant

The size of the feeding buffers is equal to 50 per cent of the total safety time of the chain of activities leading into the CPM. The calculation of all three feeding buffers is shown in Tables 8.6, 8.7 and 8.8.

Feeding buffer #1 = (Start + D + E + G + L) × 50% = **X**

Feeding buffer #1= (0 + 2 + 0 + 0 + 1) × 50% = **1.5**

Feeding buffer #1= **1.5**

Feeding buffer #2= (Start + D + I + N) × 50% = **X**

Feeding buffer #2= (0 + 2 + 4 + 0) × 50% = **3**

Feeding buffer #2= **3**

Feeding buffer #3= (Start + D + J + M) × 50% = **X**

Feeding buffer #3= (I to N) = (0 + 2 + 1) × 50% = **1.5**

Feeding buffer #3= **1.5**

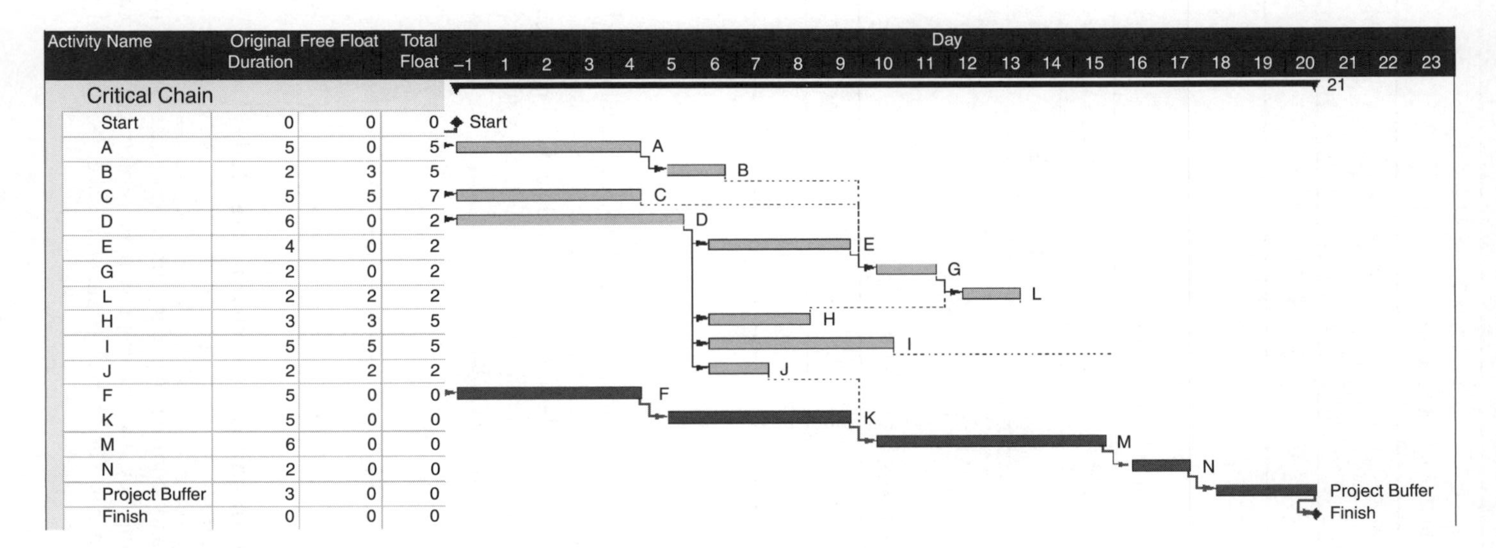

Activity Name	Original Duration	Free Float	Total Float
Start	0	0	0
A	5	0	5
B	2	3	5
C	5	5	7
D	6	0	2
E	4	0	2
G	2	0	2
L	2	2	2
H	3	3	5
I	5	5	5
J	2	2	2
F	5	0	0
K	5	0	0
M	6	0	0
N	2	0	0
Project Buffer	3	0	0
Finish	0	0	0

Figure 8.3 Part complete CC schedule with project buffer inserted.

Table 8.6 Feeding buffer #1 calculation

Non-critical chain 1	*Start* →	*D* →	*E* →	*G* →	*L* →	*Feeding buffer* →	*N*
Safety time	0	2	0	0	1	**1.5**	N/R

N/R: not relevant

Table 8.7 Feeding buffer #2 calculation

Non-critical chain 2	*Start* →	*D* →	*I* →	*Feeding buffer* →		*N*
Safety time	0	2	4	3	4.5	N/R

N/R: not relevant

Table 8.8 Feeding buffer #3 calculation

Non-critical chain 3	*Start* →	*D* →	*J* →	*Feeding buffer* →	*M*
Safety time	0	2	1	1.5	N/R

N/R: not relevant

The feeding buffers are then inserted into the CC schedule between the final non-critical activity and its succeeding critical activity, as shown in Figure 8.4.

The results for the example of the CC schedule developed using the CCS methodologies are illustrated in Tables 8.9 (contingency in the CC schedule) and 8.10 (float in non-critical chains).

Total schedule duration: **21 days**

Total time contingency in the critical chain: $0 + 0 + 0 + 0 + 0 + 3 =$ **3 days**

Percentage of safety time in the critical chain: $3/21 \times 100\% =$ **14%**

8.5.6 Critical path and critical chain case study results comparison

The total float and contingency data from the case study is shown in Table 8.11. Contingency does not represent the 'safety time' in each activity but the amount of time dedicated as an allowance to safeguard the project against risk.

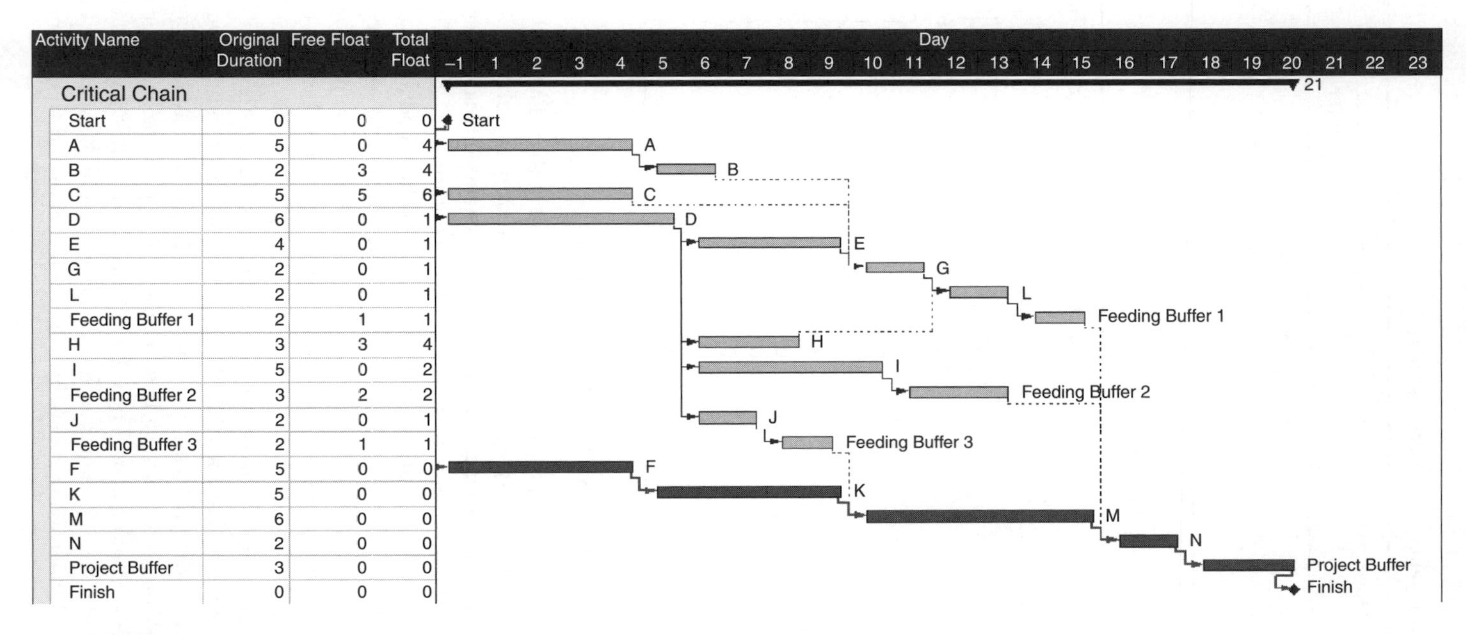

Activity Name	Original Duration	Free Float	Total Float
Critical Chain			
Start	0	0	0
A	5	0	4
B	2	3	4
C	5	5	6
D	6	0	1
E	4	0	1
G	2	0	1
L	2	0	1
Feeding Buffer 1	2	1	1
H	3	3	4
I	5	0	2
Feeding Buffer 2	3	2	2
J	2	0	1
Feeding Buffer 3	2	1	1
F	5	0	0
K	5	0	0
M	6	0	0
N	2	0	0
Project Buffer	3	0	0
Finish	0	0	0

Figure 8.4 Complete CC schedule with buffers inserted.

Table 8.9 Contingency in the critical path

Critical chain	*Start*	→ *F*	→ *K*	→ *M*	→ *N*	→	*Project buffer*	→	*Finish*
Duration	0	5	5	6	2		3		0
Time contingency	0	0	0	0	0		3		0

Table 8.10 Float in the CC schedule

Description	*Original duration*	*Safety time*	*Free float*	*Total float*
A	5	0	0	4
B	2	0	3	4
C	5	0	4	6
D	6	0	0	1
E	4	0	0	1
G	2	0	0	1
L	2	0	3	1
Feeding buffer #1	1.5	1.5	0.5	1
H	3	0	3	4
I	5	0	4	2
J	2	0	3	1
Feeding buffer #2	1.5	3	0.5	1
Feeding buffer #3	2	1.5	3	1

8.5.7 Case study discussion and conclusion

The time contingency techniques used to control risk in the CPM and CC schedules were displayed and analysed through the case study in this chapter. By testing the two scheduling techniques on a hypothetical example, results were created in the form of float, project duration and risk contingency. Table 8.11 cross compares the amount of total float in all activities of the schedules produced.

It is evident from Table 8.11 that the CPM schedule had no contingency built in (only safety time, which is not controllable) and a high amount of float in non-critical activities when compared with the float in non-critical chains of the CC schedule. The CC schedule had contingency strategically built into the 'buffers'; however, the CC schedule had less float in non-critical activities than the CPM schedule. Considering that the CCS theory expresses float as a negative aspect of a schedule due to the problems associated with Murphy's Law and Parkinson's Law (Lewis 2001), the CC schedule with less float appears to be better than the CPM schedule with more float. The CPM theory follows the view that a planner never wants to have a schedule that has no float because the risk is so extremely high that the planner won't meet the completion date and there is no time to

Table 8.11 Total float and contingency case study data comparison

Activity	*Critical path*		*Critical chain*	
	Total float	*Contingency*	*Total float*	*Contingency*
Start	0	0	0	0
A	7	0	4	0
B	7	0	4	0
C	8	0	6	0
D	2	0	1	0
E	4	0	1	0
F	0	0	0	0
G	4	0	1	0
H	7	0	2	0
I	4	0	1	0
Feeding buffer #1	N/R	N/R	1	1.5
J	2	0	1	0
Feeding buffer #2	N/R	N/R	1	3
K	0	0	0	0
L	4	0	1	0
Feeding buffer #3	N/R	N/R	1	2
M	0	0	0	0
N	0	0	0	0
Project buffer	N/R	N/R	0	3
Finish	0	0	0	0
Total duration	23		21	
Safety/CPM or CC non-critical activities	22%		14%	

N/R: not relevant

safeguard non-critical activities. The CPM schedule appears to allow more project control; however, if the values of the float and feeding buffer are added together to non-critical activities of the CC schedule, the CC schedule offers the same amount of excess time available in non-critical paths. This permits the planner to exercise the same or greater control in the CC schedule compared with the CPM schedule.

It is evident from the case study that the CC schedule provides the planner with more control over risk than a CPM schedule through the use of the CCS risk control techniques (buffers and float). One pitfall of the CCS method is that it does not give the planner as much control over the duration of individual activities as the CPM would. This makes it more difficult for the planner to effectively organise resources by activity task start and finish dates. The difficulty in locking down start and finish dates of activities in CCS is due to the fact that CCS uses 'optimistic' activity durations, which fluctuate more frequently than 'pessimistic' activity durations. Despite this management hurdle, if the CCS is integrated into the project as part of management and reporting procedures, its risk control techniques would serve to be a far better outcome for the entire project.

8.6 Summary

This chapter examined the concept of CCS, which is the implementation tool of the TOC. The traditional critical path scheduling method was reviewed and its weaknesses identified. The TOC was then briefly discussed followed by an examination of the concept of CCS with a comparison with the CPM.

Chapter 9

Multiple activity charts

9.1 Introduction

The purpose of this chapter is to examine the concept of a multiple activity chart (MAC) and demonstrate its application as a short-range resource scheduling technique.

In medium- to long-range scheduling, the work is broken down into specific activities, which are linked together to form a logical production sequence. Resources are then allocated to activities as required.

In short-range scheduling, where resources may already be committed, the work is scheduled within the bounds of committed resources. For example, in using a weekly schedule the project manager would manage and control the project from day to day around already committed resources such as cranes, hoists, other plant and labour, including subcontractors, by allocating the work to be accomplished daily to the committed resources. This approach would ensure that the committed resources are used most efficiently within the available time period.

The critical path method (CPM) is an excellent scheduling tool for developing medium- to long-range production strategies within the limits of available resources. However, it is less effective in ensuring efficiency of committed resources in short-term scheduling. A resource-driven scheduling technique such as a MAC ensures maximum efficiency of committed resources and provides graphical representation that allows management to clearly communicate work area sequencing and number of resources required by each subcontractor in the short term.

A MAC represents the most detailed form of scheduling. When required, the planner may schedule the work using very fine time-scales, such as hours or even minutes. This allows the planner to find the most effective work method that maximises efficiency of committed resources.

A MAC is highly suitable for scheduling repetitive tasks. Most construction projects are repetitive in nature. Some projects comprise a number of separate buildings or structures that are either identical or substantially similar in design, for example individual houses in a large housing project, concrete bridge piers, or cooling towers in a power-generating plant. Some other projects such as high-rise commercial buildings or freeways comprise repetitive structural elements: structural floors of high-rise commercial buildings or sections of the road surface are similar or even identical in design and are constructed in much the same manner. Developing a CPM schedule that would repeat a detailed sequence of activities from floor to floor of a high-rise building is not only tedious but often unnecessary. Techniques such as MAC and 'line of balance' (LOB) are better suited to this task. The MAC's strength is in planning and organising the work of committed resources within one repetitive cycle of work. The same planning strategy is then applied to other repetitive work cycles. A MAC is also a good visual communication aid. A LOB then schedules the work of committed resources across all repetitive cycles of work. The LOB will be discussed in Chapter 10.

While MAC is an ideal technique for short-term scheduling, its strict application in the construction industry has been sporadic. This may partially be attributed to the fact that the principles of MAC are not well known among the industry participants. However, the lack of suitable computer software for the MAC scheduling is probably the main reason for its patchy use. While simple in its format, the manual production of MAC schedules may be rather tedious, time-consuming and costly. Having said this, it is not uncommon to find that some site managers, particularly those managing repetitive tasks, will use their own hybrid version of MAC to plan out resource allocations and determine resource requirements in the short term. Despite its sporadic and modified application, MAC warrants closer examination because of the benefits it brings to scheduling.

9.2 Format of a MAC

A MAC is a two-dimensional chart. In the absence of the universal consensus on the format of this chart, its appearance may vary from country to country. In Australia, the time-scale is usually plotted on the vertical axis while committed resources are listed horizontally (see Figure 9.4). In the UK, however, the axes of the chart are usually reversed. The time-scale of a MAC would commonly be in days or hours, and may be in seconds where very detailed work cycles are scheduled, such as petroleum projects.

The format of a MAC adopted in this book is that used in Australia. The example (in Figure 9.4) includes six committed resources comprising three separate crews of formworkers, and one crew each of steelfixers, electricians and concreters. The work performed by each of the resources is then scheduled from the top of the chart to the bottom (as illustrated in Figure 9.5). In this manner, both the production restraints, given by the logic of production, and the resource restraints, given by the number of committed resources, are considered. The aim is to schedule the work in the shortest possible time while maximising the use of committed resources.

9.3 Preparation of a MAC

The first step in developing a MAC requires the planner to identify repetitive work. Some projects tend to be repetitive in their entirety, some are substantially repetitive, and others, notably sports stadiums, theatres or even hospitals, may contain no repetitive work at all.

Once repetitive cycles of work have been identified, the planner then considers alternative methods of carrying out the work. In doing so, the planner seeks methods that simplify tasks, reduce labour congestion and maximise the use of resources. These methods may be ascertained through the application of 'Work Study' as detailed in Chapter 11. The planner then evaluates available alternatives and selects a preferred method. Since efficient use of resources is essential here, in selecting a preferred method the planner may consider dividing repetitive cycles of work into smaller elements to further improve the use of resources. For example, a construction of a high-rise building comprising a typical floor slab at each level may be divided into two, three or even more segments of similar size (see Figure 9.1). The planner will consider building each segment separately in order to achieve the best possible use of committed resources. Apart from developing a construction strategy, the planner will also compile detailed information on the quantity of materials and resources that will be committed to the work. The developed planning strategy for one repetitive area of the project will then be displayed as a precedence schedule (see Figure 9.2).

In the next phase, the planner estimates the duration of repetitive cycles of work. Past experiences with similar projects and industry practice assist the planner in this task. Sometimes cycles may fall into 'natural' patterns of daily or weekly cycles. For example, a typical structural floor of a high-rise commercial building is commonly constructed on a weekly cycle. Sometimes the planner will need to estimate the duration of repetitive cycles of work from first principles. Once a construction contract is in place, however, the agreed contract period often dictates lengths of cycle times of work. For example, assume that the contract period for construction of a high rise residential building of 12 levels is, say, 12 months. Further assume that the ground works and construction of a structure up to the ground level will take four months, and the roof construction and the finishing works two months. It means that six months is available to construct 12 floors of the building, each of which would need to be completed on a two-week cycle of work.

With this information in hand, the planner determines the size of labour crews for specific trades, and specifies the type and volume of plant/equipment (see Table 9.1). The structure of a MAC schedule can now be defined in terms of resource groups that will appear in columns and an appropriate time-scale. Resource groups or columns in a MAC schedule may represent specific trades such as plumbing or formwork, and/or types of plant/equipment. They may also represent groups of activities performed by a specific resource. For example, one resource column may be defined as 'formwork to columns and slabs', while another one is 'formwork to beams and walls'.

With all the preparatory work complete, the planner is now ready to start scheduling. The best starting point is to identify key activities that signify the start and the finish of a cycle of work. These key activities are then entered in a MAC schedule. For example, construction of a repetitive slab segment of a high-rise commercial building is likely to start with the activity 'set out' and end when concrete has been poured on that segment (see Figure 9.4). These two activities signify the start and the end of one repetitive cycle. The remaining activities of the work cycle are then scheduled in between. It is unlikely that the first attempt will achieve the best possible arrangement of work and efficient use of resources. The planner may need to repeat the exercise a number of times until the best possible solution emerges. This may mean varying the capacities of committed resources, regrouping resources, changing labour crew sizes, or even varying the cycle times of work, where this is possible.

When scheduling has been completed for one cycle of work, it will be repeated for the following cycles. After about two or three work cycles have been scheduled, a regular pattern of work of committed resources is established. Since a MAC schedule is a highly visual chart, the extent of use of resources becomes immediately apparent as empty spaces in the chart (see Figure 9.5).

Before implementing a construction strategy detailed in a MAC schedule, the planner needs first to check its accuracy and adequacy. In particular, the planner needs to ensure that:

- The production logic has been maintained
- The contract requirements have been satisfied
- The resources are neither over-committed nor under-used
- Labour crew sizes are appropriate for the tasks
- Committed plant/equipment have the required capacity appropriate for the tasks.

The process and benefits of MAC in short-range scheduling is best demonstrated on a practical example, as follows.

9.4 Example of MAC scheduling

The project in question is a high-rise office building designed as a concrete frame. The building has 40 levels of office space and five levels of underground parking. The office floors are similar in size and layout between levels 1 and 38. The plan of a typical floor is given in Figure 9.1.

The contractor has been awarded a contract to construct this building in 85 weeks. The contractor's CPM schedule requires structural floors between levels 1 and 38 to be constructed in 38 weeks, which is equivalent to a one-week cycle per floor.

The contractor identifies activities relevant to the construction of a typical structural floor and allocates them to four subcontractors as follows:

- A steel-fixing subcontractor is required to tie reinforcing bars
- A formwork subcontractor is responsible for erecting and stripping formwork, setting out floors, installing construction joints and trimming formwork

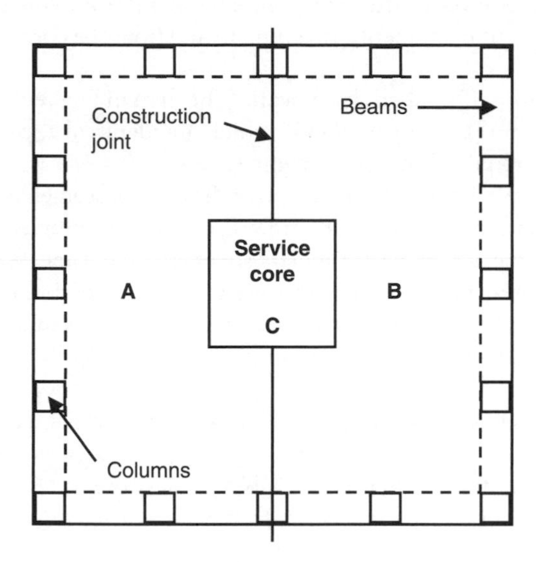

Figure 9.1 A floor plan of the case study project.

- A concreter subcontractor is responsible for placing concrete
- An electrical contractor is required to install in-slab services.

A contractor will erect one tower crane on the site that will service the four subcontractors.

The contractor's planner then develops a construction strategy to meet the required cycle time of one week and ensure maximum use of committed resources. It is unlikely that the latter objective will be achieved if each structural floor is built in one piece. But if these are divided into two or more segments and each segment is built separately, it may be possible to achieve continuity of the subcontractors' work by moving them from segment to segment.

Let's assume that the planner divides the structural floor into three segments A, B and C, where segments A and B are the mirror-image halves of the floor and the segment C is the construction of the core (Figure 9.1).

The planner decides to build slabs, columns and beams using conventional formwork, but selects slipforming for construction of the core. Slipforming is a well-known construction method of building concrete structures by continuously raising steel formwork during a concrete pour. The form is assembled at ground level and stays in one piece until the core has been completed. It climbs using its own jacking system. The planner intends to keep building the core at least four levels above the structural slabs.

After the evaluation of alternatives, the planner has adopted a strategy that requires construction of segment A to start first, followed by segment C and ending with segment B on each level of the structure. The planner intends to move subcontractors from segment to segment, including to and from the core, to maximise continuity of work.

The planner has prepared two CPM schedules of work. The first, in Figure 9.2, shows a construction sequence for segments A and B, which are identical except for a construction joint that is required only in segment A.

The second schedule, in Figure 9.3, shows construction of the core for segment C.

Next, the planner has determined appropriate crew sizes for the volume of work to be undertaken from the work content summary sheet given in Table 9.1.

This information helps the planner to determine what activities will be performed by what resources, and the size of labour crew for each such resource. The planner's decision is to use seven resource crews, each with the following work tasks:

- Crew 1, of two persons: set out, trim formwork, construction joints, core blockouts
- Crew 2, of eight persons: reinforce columns, slabs, beams and core
- Crew 3, of four persons: form columns, strip formwork, clean slipform formwork
- Crew 4, of four persons: form beams, trim formwork, strip formwork
- Crew 5, of four persons: form slab, strip formwork
- Crew 6, of two persons: install electrical services in slabs and core
- Crew 7, unspecified: pour concrete.

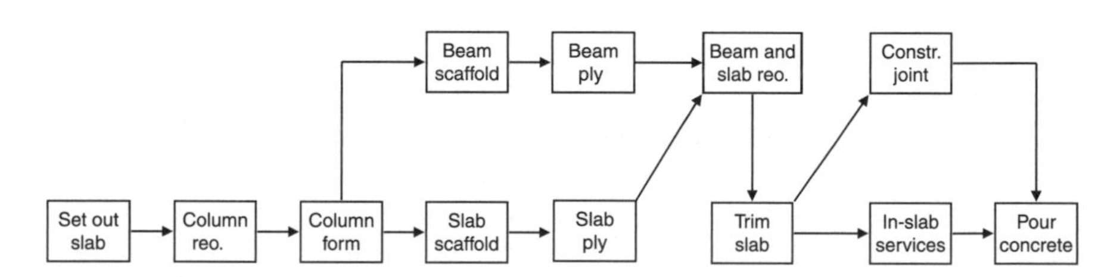

Figure 9.2 A schedule of work for segments A and B.

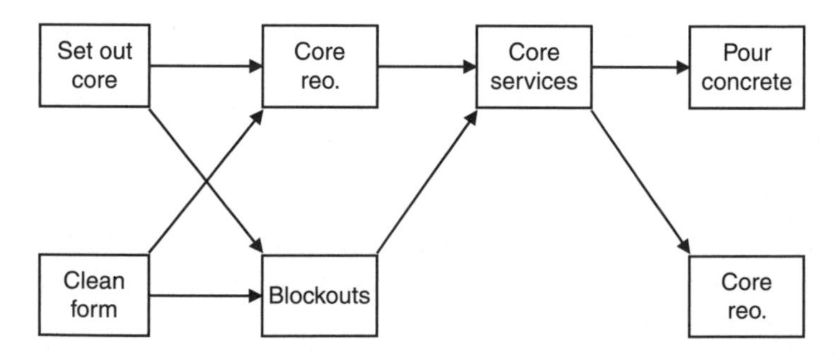

Figure 9.3 A schedule of work for segment C.

Table 9.1 The work content summary sheet for the project in question

Work content summary sheet		
Slab		
Reinforcement		
Columns	32 person-hours	4 persons × 8 hours
Beams and slab	96 person-hours	4 persons × 24 hours
Formwork		
Columns	48 person-hours	4 persons × 12 hours
Slab		
scaffolding	32 person-hours	4 persons × 8 hours
slab plywood	32 person-hours	4 persons × 8 hours
Beams	64 person-hours	4 persons × 16 hours
Stripping		
plywood	64 person-hours	4 persons × 16 hours
scaffolding	16 person-hours	4 persons × 4 hours
Miscellaneous		
Set out	8 person-hours	2 persons × 4 hours
In-slab services	16 person-hours	2 persons × 8 hours
Construction joint	8 person-hours	2 persons × 4 hours
Trim slab, beam	16 person-hours	2 persons × 8 hours
Core		
Reinforcement		
	96 person-hours	4 persons × 24 hours
Miscellaneous		
Set out	8 person-hours	2 persons × 4 hours
Services	16 person-hours	2 persons × 8 hours
Blockouts	40 person-hours	2 persons × 20 hours
Clean form	32 person-hours	4 persons × 8 hours
Concrete		
One pour per day		

The first five crews are expected to stay on the job permanently until the structure has been finished. Because crew 6, 'Electrical services', performs one activity only per segment, its full use cannot be achieved and it is likely that it will be utilised for other activities of the project. The work of crew 7, 'Placing of concrete', does not require continuity of work. Common work practice is to give sufficiently long notice to concrete placers to come to the site to place concrete as a one-off activity.

Let's assume that one production week is equal to six working days. Let's further assume that 'Stripping of formwork' takes place three floors below the working floor. It has already been decided that slipforming of the core will take place four levels above the working floor.

A MAC schedule with seven resource columns can now be prepared. Its timescale is in days and hours. To ensure that a six-day cycle of work is achieved, the planner initially schedules the first and the last activity in each segment per cycle. These are the key activities. For the project in question, the first activity of each segment is 'Set out' and the last 'Pour concrete'. Clearly, to meet the project's time objective, concreting of individual segments needs to take place every second day. This is illustrated in Figure 9.4, which provides a template of

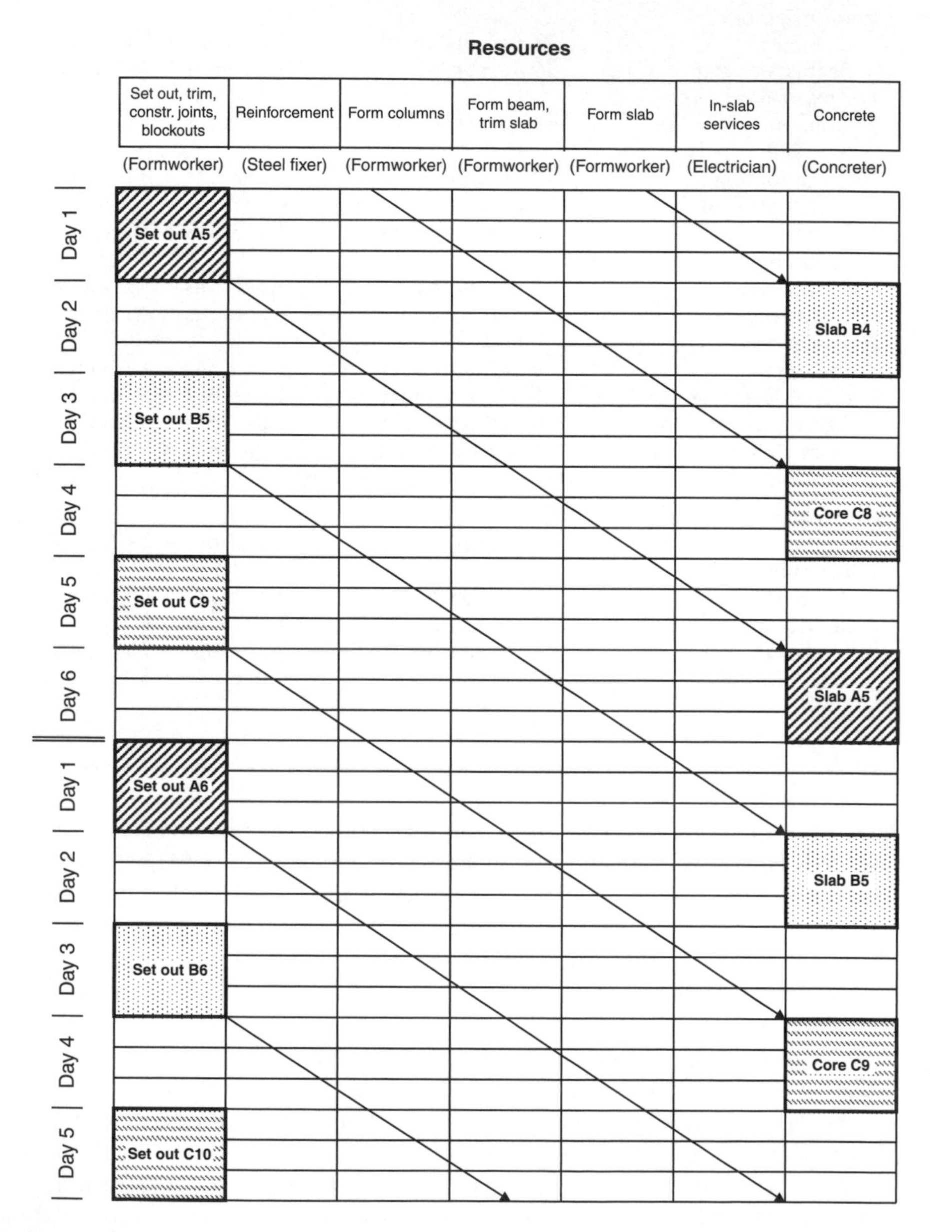

Figure 9.4 A MAC schedule with the key activities.

key activities in each segment. The pattern of work indicates that provided the remaining activities can be fitted in, all three structural floor segments will be completed in six days.

The remaining activities can now be scheduled between the key activities. It should be noted that the scheduling process using a MAC should not start on level 1, the first typical floor, because its construction may take a little longer than on other typical floors. This is due to a 'learning curve' syndrome, where resources do not achieve full productivity immediately. Common practice is to start a few levels higher, where a regular cycle time of work is expected to be constant and then schedule the first few floors by using the experience already gained. In this example, let's begin scheduling on level 5.

In scheduling individual activities within the constraints of time, committed resources and the defined logical pattern of work is not easy. The work is laborious and time-consuming, and may require a number of trials. One possible solution is illustrated in Figure 9.5. It shows that the work associated with building a typical structural floor can be completed within six days and with excellent use of committed resources. Please note that the nominated crew of eight workers in the resource columns 'Reinforcement' was actually split into two crews, of four workers each, to meet the work demand. A similar split in the resource crew occurred in the resource column 'Form beam, trim slab', but only on days 1 and 6 of the cycle.

There may be plausible alternative solutions and the reader is encouraged to seek them out.

As the structure rises, the increasing wind velocity is likely to adversely affect the crane's speed and efficiency. The length of vertical movement of materials and people is also going to increase progressively, possibly adding extra time to the cycle of work.

Since the work on levels 1, 2, 3 and 4 is likely to take longer than six days, the planner will compensate for the learning curve syndrome by adding progressively more time to the cycle of work for lower floors. Similarly, the planner will need to add more time to the work cycle for the last few floors, which are unlikely to be of the same design as the typical floor.

The arrangement of work shown in the MAC schedule in Figure 9.5 achieves the required performance goal of a six-day cycle per typical floor. Because MAC schedules are resource-based, they are largely intolerant of delays to activities. However, since the risk of delays during construction is commonly high, the planner will need to implement a number of measures to ensure integrity of the MAC schedule. These may include:

- Adding an appropriate time contingency to each activity or each cycle of work
- Having additional resources on standby
- Maintaining adequate supervision of the work
- Committing to effective process of monitoring and control.

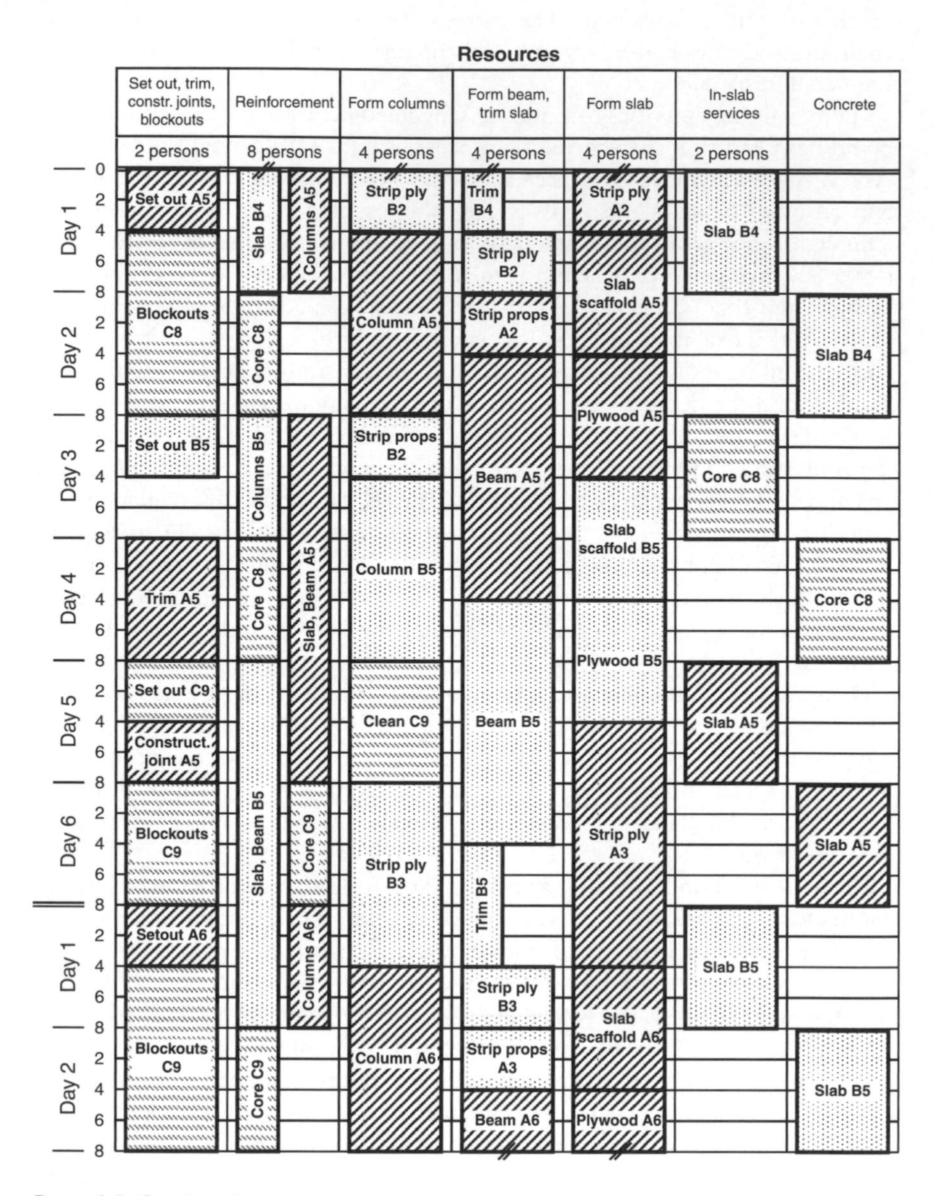

Figure 9.5 Continued opposite.

Specific scheduling outcomes achieved by using a MAC must be integrated with the overall CPM project schedule in order to maintain the planned flow of work and the provision of resources across the entire project. It is particularly important to ensure that materials are available in the required quantities and can be delivered in the specified time-frame, that the materials-handling equipment has

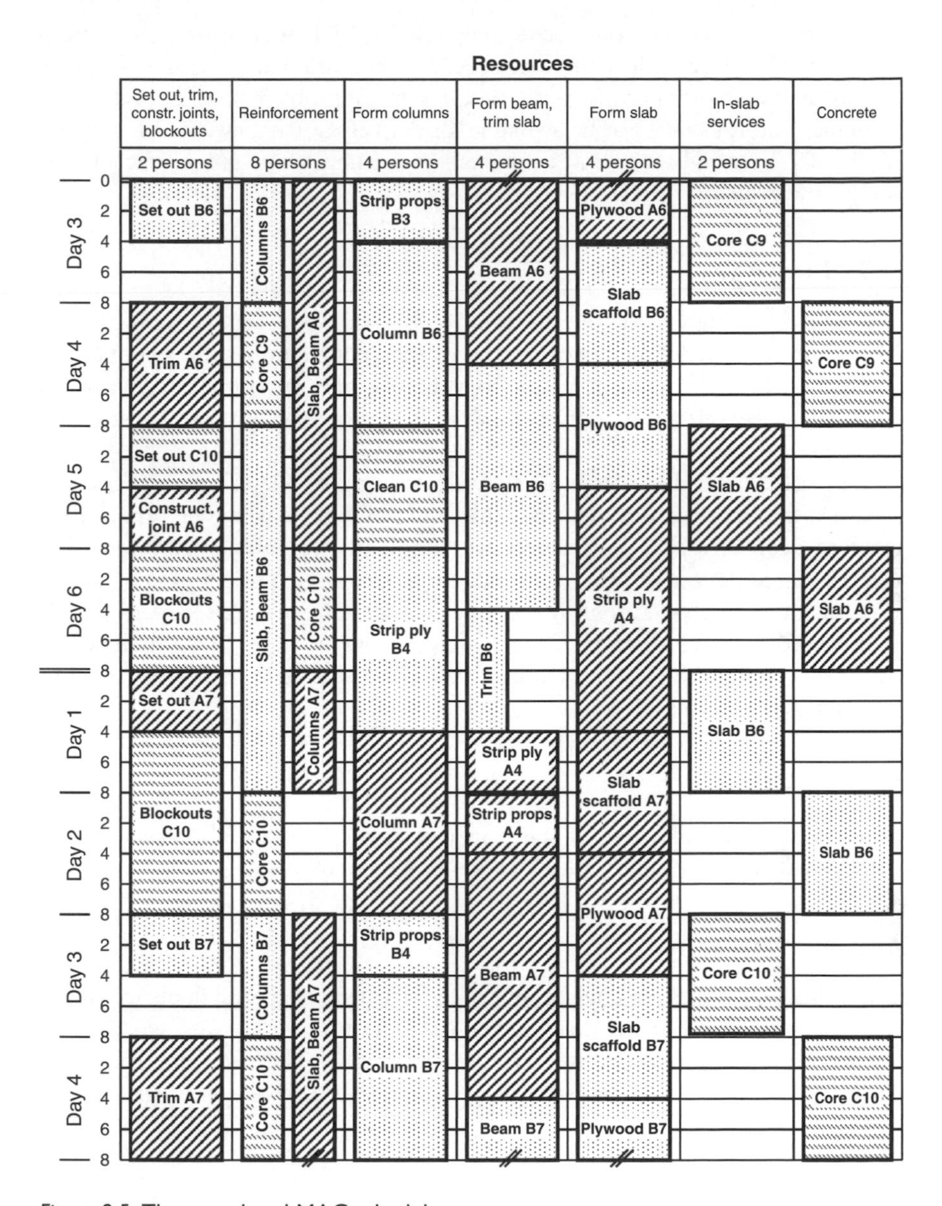

Figure 9.5 The completed MAC schedule.

sufficient capacity to service project activities, that the required labour resources are available, and that site amenities are adequate for the required labour force.

Let's assume that the planner is able to reduce the cycle time of work related to the construction of structural floors from six to five days per floor by using the available resources better. In theory, the planner would expect a significant reduction in the overall project schedule without unduly increasing the construction cost. However, unless the extra speed of building the structure is closely coordinated with other project activities, significant problems could develop. For example, materials may not be available for delivery at the new rate of demand, or the site materials-handling equipment may simply run out of capacity to service the project.

9.5 Summary

The MAC is a highly effective short-range scheduling technique. It is particularly suited for scheduling repetitive work. Because it examines the work tasks at the highest level of detail, it is able to assist planners and project managers in developing effective production methods that achieve efficient use of committed resources.

The lack of suitable computer software to support MAC severely restricts its use in the construction industry. This may well be due to the fact that MAC has only been seen as a short-term scheduling method. Some might argue that developing computer software for scheduling very short project periods makes no economic sense given that by the time a computer-generated MAC schedule is produced, checked and finalised it may be out of date. Nevertheless, the fact remains that the befits of using MAC in short-term scheduling cannot be ignored.

Exercises

Solutions to the following exercises can be found on the following website: http://www.routledge.com/books/details/9780415601696/

Exercise 9.1

Three different trades are engaged in the construction of typical floors in a high-rise building. The work schedule is given in Figure 9.6.

Using a MAC, arrange the activities in the schedule so that the three trades are employed most productively. Assume that only one crew per trade is available.

Exercise 9.2

The activities in Table 9.2 are associated with the installation of suspended ceilings on typical floors in a high-rise office building.

The area of a typical floor is 30 m × 20 m. Suspended ceilings are built to a 1.0 m × 0.5 m grid.

Determine appropriate crew sizes for the four crews of workers and then prepare a schedule of work using a MAC. Ensure that the resources are used efficiently.

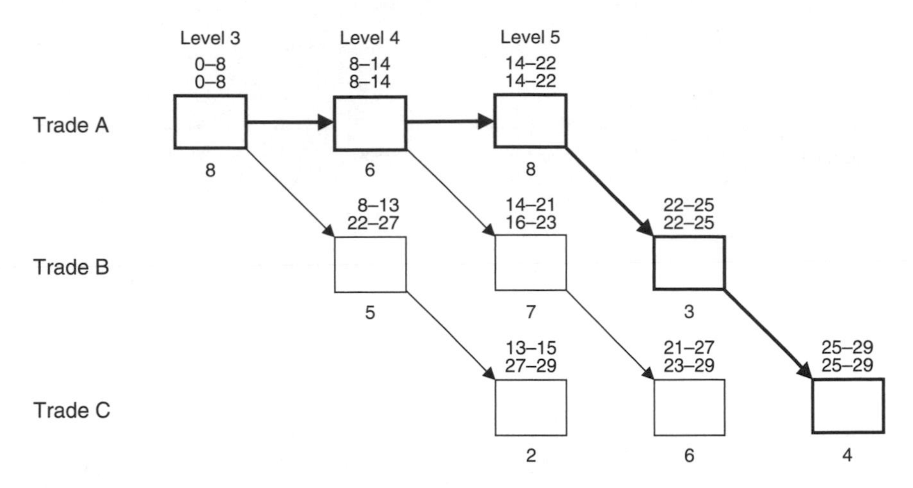

Figure 9.6 The precedence schedule for the example.

Table 9.2 Data for the MAC example

Crews	*Activities*	*Planned production rates*
Ceiling crew	Set out ceiling grid	8 person-hours
	Fix ceiling hangers @ 2 m centres	0.1 person-hour/hanger
	Fix ceiling framing	50 linear m/person-hour
	Fix ceiling tiles	12.2 m2/person-hour
Sprinkler head crew	Fix sprinkler heads @ 3 m centres to previously installed sprinkler pipes	3 heads/person-hour
Lights fitting crew	Fit light fittings into ceiling grid @ 3 m centres	6 fittings/person-hour
A/C register fitting crew	Air conditioning registers @ 4 m centres	3 registers/person-hour

A construction sequence of activities involved in the installation of ceilings is given in the precedence schedule in Figure 9.7.

Exercise 9.3

A crane is being used to lift pallets of concrete bricks into the building. One crew of labourers unloads concrete bricks from the truck onto pallets and the other unloads pallets in the building. These crews of labourers are also responsible for hooking and unhooking pallets to the crane. One set of slings only is available for use. Table 9.3 gives a list of activities and their standard times expressed in minutes.

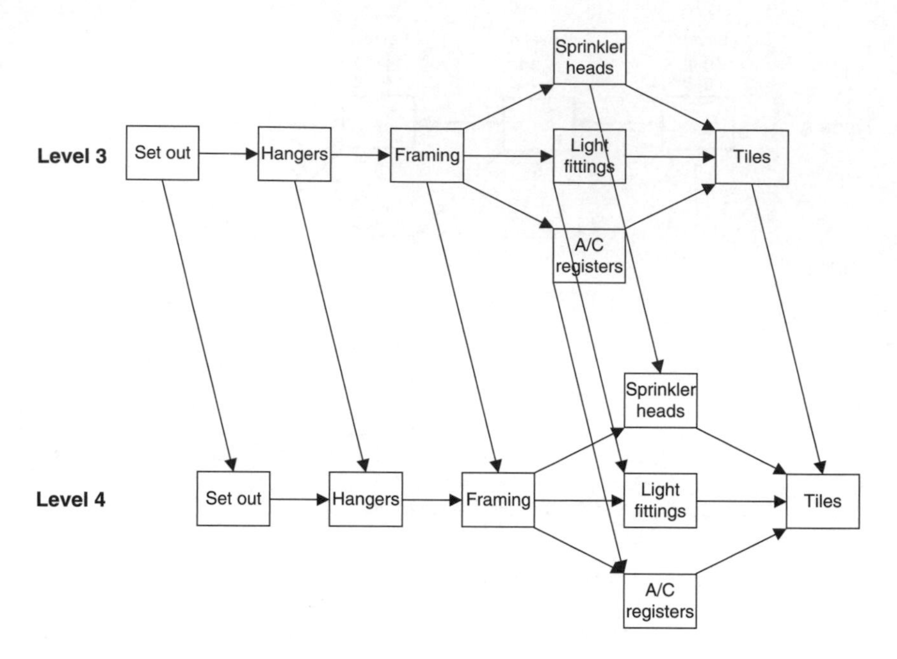

Figure 9.7 The precedence schedule for the example.

Table 9.3 Data for the MAC example

Activities	*Standard times (minutes)*
Remove bricks from truck and fill one pallet	6
Lift full pallet into building with crane	2
Unload pallet in building	5
Lower empty pallet with crane	1
Hook one pallet onto crane	0.5
Unhook one pallet from crane	0.5
Lift crane hook (without pallet)	2
Lower crane hook (without pallet)	1

There are three pallets available, each holding 100 bricks. They are located on the ground floor adjacent to the 'unloading area' for trucks. One truck can hold 800 bricks.

Using a MAC, illustrate a method of getting bricks into the building. How long will it take to unload the first truck? What is the cycle time of work measured for three pallets?

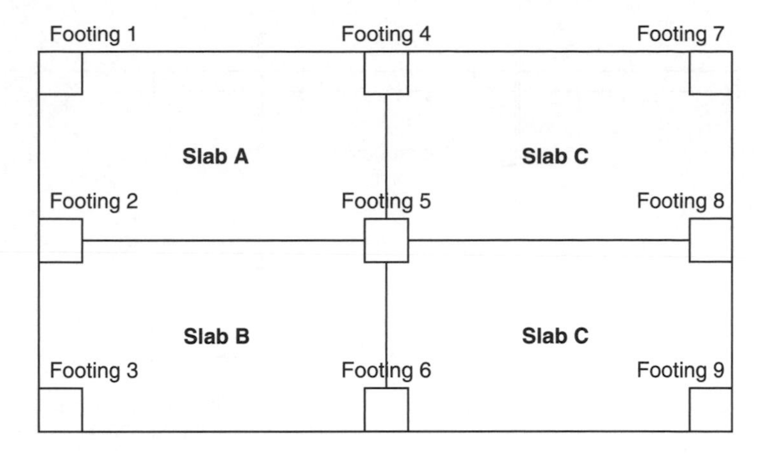

Figure 9.8 The footing plant

Table 9.4 Data for the MAC example

Crews	*Activities*	*Time (hours)*
Excavator	Excavate footings	3 each
	Trim slab areas before reinforcing	2 each
Reo. fixer	Reinforce footings	2 each
	Reinforce slab areas	1 each
Concreter	Pour footings	1 each
	Pour slab areas	2 each

Exercise 9.4

A footing plan for a building project is shown in Figure 9.8. Specific activities, durations and crews are given in Table 9.4.

Prepare a MAC schedule showing the most efficient method of building the footings.

Exercise 9.5

Figure 9.9 shows the layout of a typical floor in a high-rise concrete frame building. Prepare a short-range MAC schedule for construction of the structure (slabs, columns, walls, beams and stairs) and some miscellaneous items detailed in the 'Work content summary sheet'.

The structural engineer has provided for two construction joints in the floor structure. These may be conveniently used for breaking up the floor into three segments A, B and C.

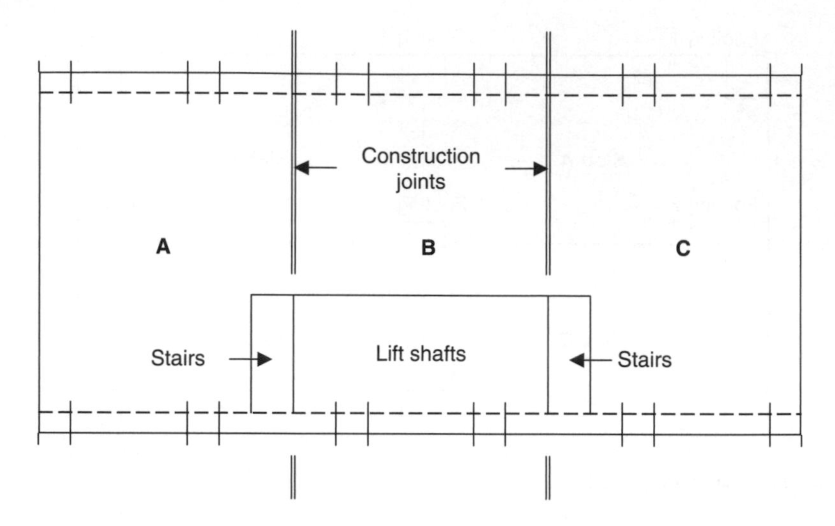

Figure 9.9 The floor plan.

Segment C is identical to A except for a construction joint. Segment B is similar to segments A and C, with a somewhat smaller floor area but larger wall area. For the purpose of this question, assume that the construction sequence for each segment is largely the same. A precedence schedule of the construction sequence for one segment is given in Figure 9.10.

The activities related to the construction of a typical floor together with resource demand levels are given in the 'Work content summary sheet' in Table 9.5.

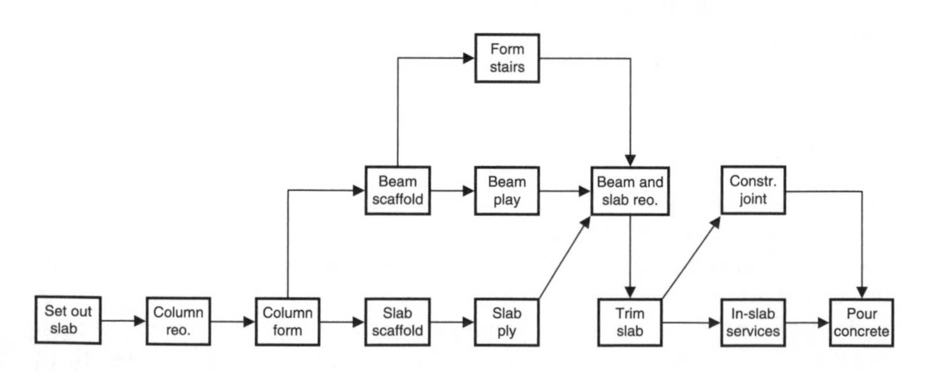

Figure 9.10 The precedence schedule.

Table 9.5 Data for the MAC example

Work content summary sheet (for one segment)		
Reinforcement		
Columns	32 person-hours	4 persons × 8 hours or 8 persons × 4 hours
Beams and slabs	96 person-hours	4 persons × 24 hours or 8 persons × 12 hours
Formwork		
Columns	70 person-hours	7 persons × 10 hours
Stairs/walls	42 person-hours	7 persons × 6 hours
Slabs		
scaffolding	24 person-hours	3 persons × 8 hours
slab plywood	24 person-hours	3 persons × 8 hours
Beams	128 person-hours	8 persons × 16 hours or 4 persons × 32 hours
Miscellaneous		
Setout	8 person-hours	2 persons × 4 hours
In slab services	16 person-hours	2 persons × 8 hours
Constr. joint	8 person-hours	2 persons × 4 hours
Trim slab, beam	16 person-hours	2 persons × 8 hours
Concrete	One pour per day	

Note: Stripping of formwork is not considered.

Chapter 10

The line of balance technique

10.1 Introduction

This chapter will overview the line of balance technique (LOB) and examine its capacity to schedule repetitive construction tasks.

The repetitive nature of construction projects was already discussed in Chapter 9 in relation to the technique of a multiple activity chart (MAC). It was explained that construction projects such as large housing estates, high-rise commercial buildings, highways, tunnels, pipelines, bridges and the like consist of numerous, highly repetitive elements. Such projects are commonly referred to as repetitive or linear construction projects. It was also suggested that a critical path method (CPM) technique is less appropriate for scheduling repetitive projects. Other techniques such as MACs and LOB are preferred. A MAC was described as a highly effective short-range planning technique, particularly suited for scheduling repetitive tasks. LOB is effective in medium- to long-range scheduling.

Mattila and Abraham (1998) and Al-Harbi *et al.* (1996) described techniques that may be used for scheduling repetitive projects:

- CPM
- Optimisation methods such as the dynamic programming model, the linear scheduling model (LSMh) and the repetitive project modelling (RPM)
- Graphic techniques such as the LOB, the vertical production method (VPM) and the linear scheduling method (LSM).

Scheduling of repetitive tasks using a CPM method results in schedules with a very large number of activities. Their visualisation and interpretation is often difficult. Optimisation methods treat scheduling of repetitive projects as a dynamic process. They look for efficient solutions in terms of cost and time while maintaining the constraints related to the production rate and continuity of work of committed resources. Optimisation methods will not be discussed in this book.

The graphic methods plot repetitive activities as diagonal lines on an X–Y graph with time on the horizontal axis and location on the vertical axis. The slopes of the lines represent production rates of the activities. LOB, VPM and LSM techniques are very similar and are often treated as being the same. VPM (O'Brien 1975) is best suited to scheduling repetitive activities in high-rise building projects while LSM (Mattila and Abraham 1998) is more applicable to truly linear construction projects where activities are repetitive for the entire project duration, such as pipeline or highway construction.

LOB embraces the features of both VPM and LSM, and in addition creates a control graph from which the planner can determine how many repetitive segments and sub-segments will be completed at specific times.

A technique of LOB will now be examined in detail and its link to MAC emphasised. First, the main features of LOB will be explained. The concept of delivery programme in LOB will then be introduced, after which a process for developing a LOB schedule with single and multiple crews will be discussed. While only a manual approach to LOB scheduling will be discussed in this chapter, the reader should be aware of various LOB software that has emerged in recent years including Graphisoft Control 2005 and Q Scheduling.

10.2 Concept of LOB

LOB is a relatively simple technique suitable for medium- to long-range scheduling of repetitive projects. It is a resource-based scheduling technique with the prime concern of ensuring continuity of work and efficient use of committed resources.

LOB relies on information generated by a MAC. The task of a MAC is to organise the work at the highest possible level of detail in order to determine work cycles of repetitive processes, and to ensure the most efficient use of resources in such repetitive cycles of work. LOB then links repetitive cycles of work together to form an overall schedule.

Construction of repetitive projects is normally carried out by sequencing discrete activities, which are identical in each repetitive segment. For example, a large housing project comprises many similar or identical houses. Construction of one house is commonly broken up into activities or groups of activities referred to as packages. Crews of workers perform the tasks associated with activities or packages. The project is then constructed by arranging the work of each crew to proceed continuously and sequentially from the first to the last house. A MAC assists in defining sizes of crews for each activity or package. Sometimes one crew per activity or package is sufficient, sometimes multiple crews may be required to achieve the desired rate of progress.

If individual activities or packages in each repetitive segment have the same rate of progress, they are arranged to progress one after the other from the first to the last house. This is illustrated in Figure 10.1. A gap between individual activities is a buffer zone, which serves as a time contingency of each activity.

A uniform rate of progress of individual activities is rarely achieved. A more realistic scenario is to expect rates of progress to vary from activity to activity.

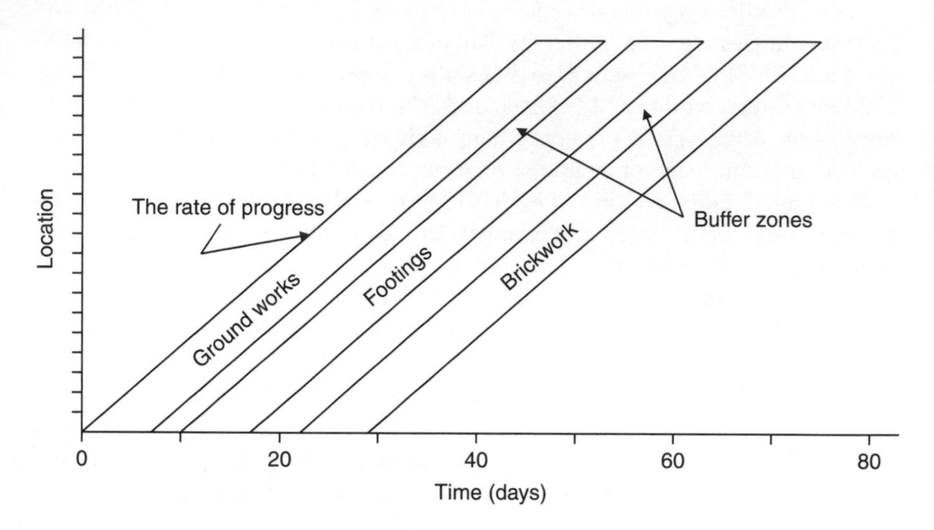

Figure 10.1 A LOB schedule of activities with identical cycle times.

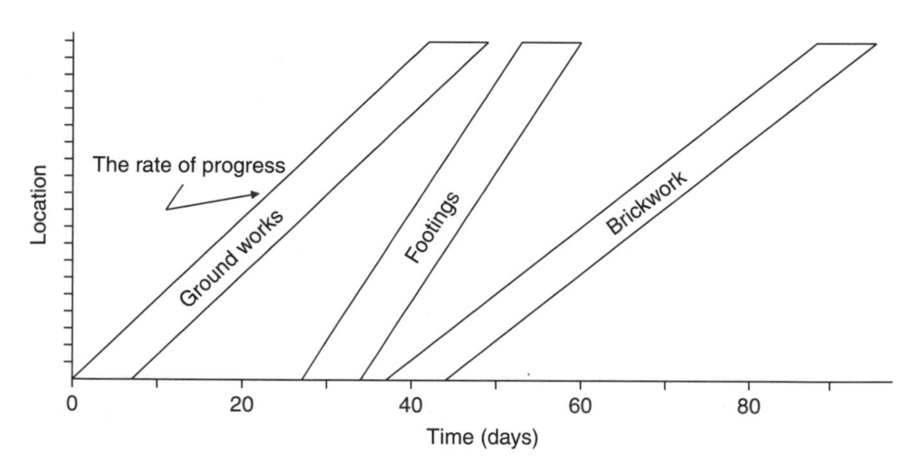

Figure 10.2 A LOB schedule of activities with different cycle times.

Activities of such varying rates of progress are then linked together so that continuity of work of each crew is maintained. This is illustrated in Figure 10.2. The rate of progress of activity 'Footings' is faster than the rate of progress of the preceding activity 'Ground works'. To avoid the possibility of discontinuity of work in activity 'Footings', the planner will schedule this activity from the finish of the preceding activity 'Ground works' by, first, determining the completion date of the last 'Footings' activity, which will occur immediately after completion of the last 'Ground works' activity, and second, determining the start of the first 'Footings' activity.

Conversely, because the rate of progress of activity 'Brickwork' is slower than the rate of progress of the preceding activity 'Footings', by starting the first activity 'Brickwork' immediately after completion of the first activity 'Footings', there is no possibility of discontinuity of work in activity 'Brickwork'.

The rate of progress of each activity or package may or may not remain constant over the project period. For example, pipeline construction may progress at a constant rate if the ground conditions are unchanged but may increase or decrease with the change in ground conditions. In constructing high-rise buildings, cycle times of work from floor to floor are likely to increase in duration as the structure rises. This is largely due to the extra time needed to lift materials and transport people to the rising structure, and possibly also to the increasing intensity of wind.

10.3 Concept of delivery programme in LOB

The rate of progress of repetitive projects is often controlled by a contract. For example, assume that the contract requires the contractor to deliver at least two fully completed houses each week over the contract period of 50 weeks. Let's further assume that there are in total 100 houses to be constructed under this contract and that the first house will take ten weeks to complete. It means that the remaining 99 houses will need to be delivered between weeks 10 and 50, which is equivalent to approximately 2.5 houses delivered per week. This is illustrated in Figure 10.3. The contractor would then need to develop a construction strategy that would meet the required delivery rate of houses. This will be addressed later in the chapter.

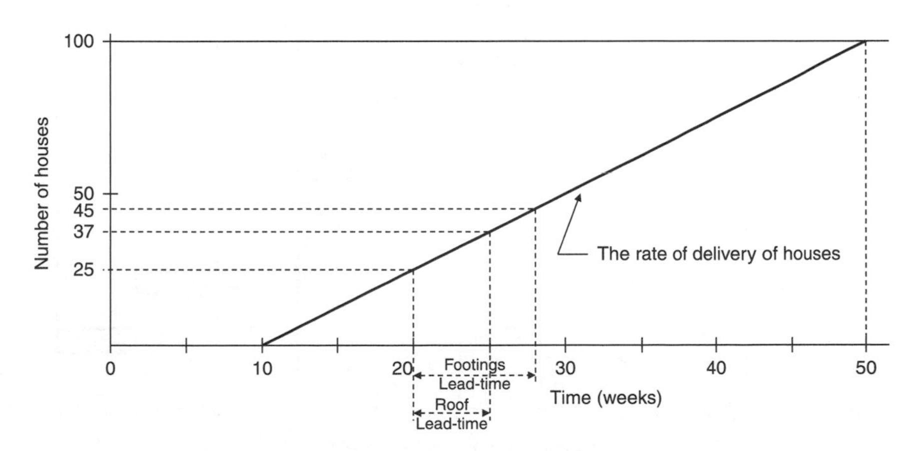

Figure 10.3 The delivery rate graph.

Having established the required delivery rate of completed houses, it is possible to use the delivery rate graph to establish, at specific intervals, the planned production output of houses and their individual activities. As construction gets under way, the planned volume of output will be used as a control tool against which the actual production will be assessed. But first, let's define the concept of 'lead-time'.

The concept of 'lead-time' helps in determining the number of houses and individual activities that will need to be completed at specific intervals. It is a period (in weeks, days or other time-units) by which a particular activity must precede the end activity if the delivery of a repetitive unit is to be made on time.

The bar chart schedule in Figure 10.4 illustrates the concept of 'lead-time'. The schedule shows a sequence of activities related to building a house in ten weeks. When the time-scale is reversed at the foot of the bar chart, lead-times for each activity can be determined. For example, the activity 'Footings' has eight weeks of lead-time. It means that when this activity is completed, it will take another eight weeks to complete the house. Similarly, the activity 'Clean-up' has zero lead-time, which suggests that when it is completed the house is fully completed.

Let's use lead-times to plan production output levels. For example, how many houses and their individual activities will be completed on week 20 using the rate

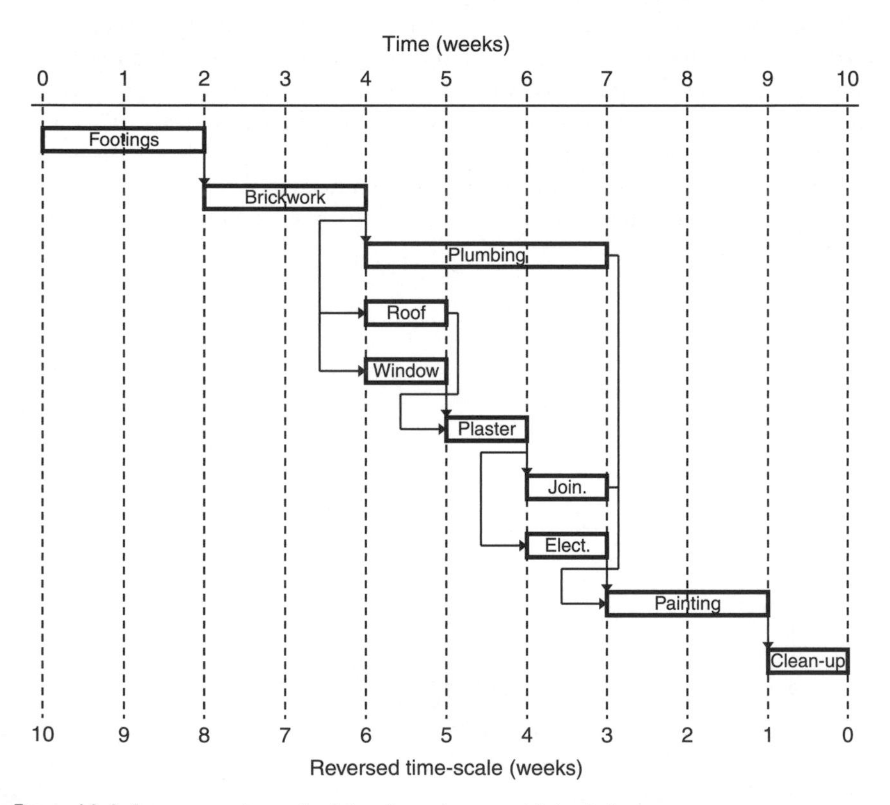

Figure 10.4 A construction schedule of one house with lead-times.

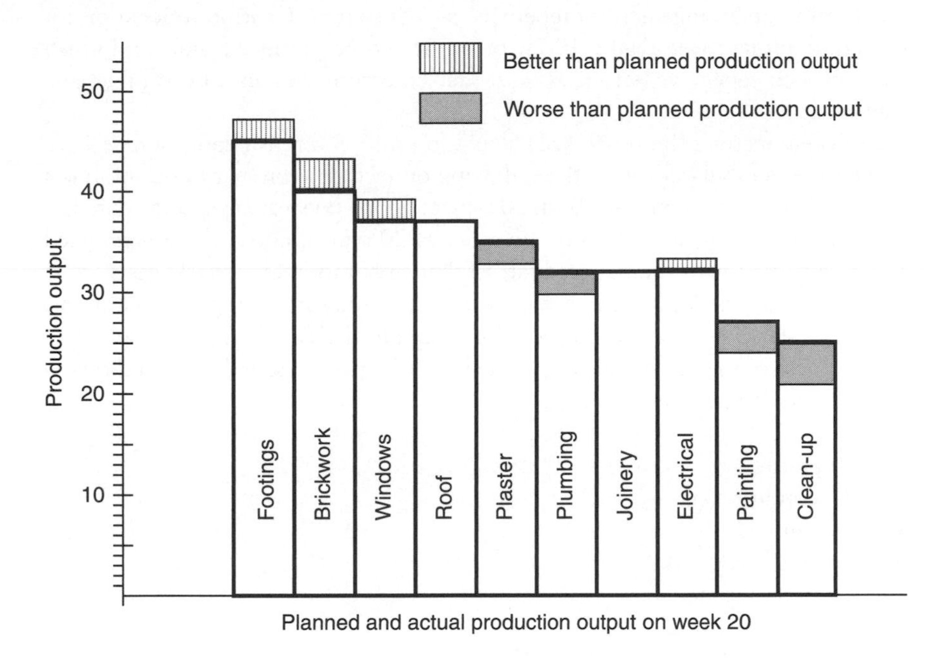

Figure 10.5 The planned and actual production output graph.

of delivery graph in Figure 10.3? When the last activity 'Clean-up' is completed, lead-time is zero, which is the lead-time of a fully completed house. A vertical line drawn at week 20 intercepts the rate of delivery line. The intersection point represents the number of completed houses, which is read off of the vertical axis as 25. The volume of completed activities on week 20 are determined in much the same manner. Let's take activity 'Footings' first. Because its lead-time is eight weeks, a vertical line is drawn eight weeks to the right from week 20, which is week 28. The interception point on the rate of delivery line corresponds to 45 completed footings, read off of the vertical axis. Similarly, since activity 'Roof' has a lead-time of five weeks, a vertical line is drawn at week 20 + 5 = 25 and corresponds to 37 completed 'Roof' activities on week 20. This process is then repeated for the remaining house activities. The planned production output of each activity on week 20 is presented graphically in Figure 10.5. The actual production output is then marked up on the graph. In this format, the difference between the planned and actual production output is clearly apparent. Similar graphs may be prepared at some other time intervals.

10.4 Developing a LOB schedule

The fundamental aspects of preparing a LOB schedule were already discussed in section 10.2 and illustrated in Figures 10.1 and 10.2. A LOB schedule shows graphically the arrangement of repetitive activities from location to location for the entire project (see Figure 10.7). Before it can be produced, start and finish dates of such repetitive activities must first be determined in a LOB table (see Table 10.1).

Let's demonstrate the process of calculating start and finish dates of repetitive activities in a tabular form for the following project. The project in question is a high-rise commercial building with 20 typical floors. Because no specific date has been given for the completion of the fitout of 20 typical floors, there is no need to be concerned with the required rate of delivery of individual packages.

The building is designed as a concrete frame structure, comprising columns, walls and slabs. For the purpose of this example, LOB will not be applied to construction of the 20 typical floors. It will only be used to schedule the following activities/trade packages:

- Stripform
- Air conditioning
- Hydraulics
- Internal brickwork
- Concrete facade
- Suspended ceiling
- Interior finish.

The work in each trade packages will be performed by one crew of subcontractors. It is assumed that construction of structural floors proceeds well in advance of the above activities. A precedence schedule of work necessary to fully complete each typical floor is given in Figure 10.6.

Table 10.1 shows in the first two columns a list of trade packages and their respective cycle times. Assume that they have been determined by a MAC.

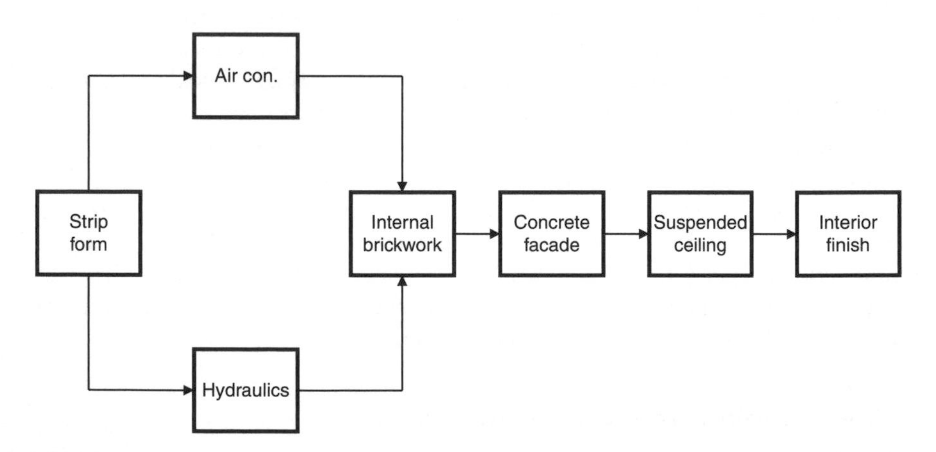

Figure 10.6 A precedence schedule of work related to one typical floor.

Table 10.1 The LOB table of individual trade packages

Trade packages	*Duration per cycle of work (days)*	*Total duration (days)*	*Start of work package (days)*	*Finish of work package (days)*
1. Strip formwork	6	120	0	120
2. Air conditioning ducts	8	160	6	166
3. Hydraulics	5	100	25	125
4. Internal brickwork	7	140	33	173
5. Concrete facade	5	100	78	178
6. Suspended ceiling	6	120	83	203
7. Interior finishes	8	160	89	249

The total duration of the work associated with each repetitive trade package over 20 typical floors is calculated in the third column. The table shows that cycle times of work of the individual packages vary between five and eight days. To ensure continuity of work of subcontractors, a succeeding package that has a slower or the same rate of progress than its immediately preceding package will be scheduled as soon as the first cycle of work of its preceding package has been completed. If everything goes according to the plan, this succeeding package will not interfere with the work of the preceding one. However, if a succeeding package has a faster rate of progress than the preceding package, its start would need to be delayed sufficiently to prevent interference with the preceding package.

Let's now calculate start and finish dates of the trade packages in columns 4 and 5. 'Strip formwork' is the first package in the production cycle. It starts on day 0. Since its total duration is 120 days, it will finish on day 120.

The start of the next two trade packages, 'Air conditioning ducts' and 'Hydraulics', is influenced by the completion of the preceding package, 'Strip formwork'. Let's schedule the package 'Air conditioning ducts' first. Because its duration per cycle of work is longer than that of the preceding package 'Strip formwork', it can be scheduled to start as soon as the first 'Strip formwork' activity has been completed, which will be on day 6. This package will be fully completed on day 166.

The cycle of work of the package 'Hydraulics' is faster than that of the preceding package, 'Strip formwork'. It means that it cannot be scheduled to commence as soon as the first 'Strip formwork' activity has been completed because it would continually interfere with it and result in discontinuity of work. Rather it will be scheduled from the finish date of the last 'Strip formwork' activity. This last 'Strip formwork' activity is scheduled for completion on day 120 and therefore the last 'Hydraulics' activity will be completed five days later on day 125. The first 'Hydraulics' will then start on day 25. Its progress will be uninterrupted and it will not interfere with its preceding activity.

The next package to schedule is 'Internal brickwork'. Its start is dependent on completion of two preceding pack ages, 'Air conditioning' and 'Hydraulics'. Let's consider these two preceding packages one at a time. With regard to 'Air conditioning', the 'Internal brickwork' package is faster per cycle of work. It will therefore be scheduled from the finish of the last 'Air conditioning' activity. Because the last 'Air conditioning' activity is scheduled for completion on day 166, the last 'Internal brickwork' activity will be finished seven days later, on day 173. The start date of the 'Internal brickwork' package will then be day 33. With regard to 'Hydraulics', 'Internal brickwork' is slower per cycle of work and will therefore start as soon as the first 'Hydraulics' activity has been completed, which is on day 30. The package 'Internal brickwork' will then be completed on day 170, which is three days before its other preceding package, 'Air conditioning', would be fully completed. Clearly, the start and finish of the package 'Internal brickwork' is governed by its preceding package, 'Air conditioning' and not by 'Hydraulics'.

The start and finish dates of the remaining trade packages are determined in the same manner. The first fully completed floor will be ready on day 97, which is the finish date of the first 'Interior finishes' package. The delivery rate of the remaining floors is one every eight days thereafter (this rate is equal to the cycle time of the last package, 'Interior finishes'). The total project duration is 249 days. The first fully complete floor will be delivered on day 97, which is the start of the first 'Interior finishes' activity plus its duration. Therefore, 89 + 8 = 97.

The trade packages have been scheduled to ensure continuity of work from start to finish. Because unforeseen risk events may cause delays in some activities, the prudent planner will safeguard against delays by adding time contingency to the schedule. This is commonly done in the form of a buffer zone inserted between trade packages. In Table 10.2 and Figure 10.7 buffer zones of six days between each package have been inserted into the schedule. The project would be completed on day 279 with the first fully completed floor ready on day 127.

In the LOB graph, the impact on project period by slow packages is clearly apparent. If it is possible, for example, to reduce the cycle time of work of the

Table 10.2 The LOB table with buffer zones of six days

Trade packages	*Duration per cycle of work (days)*	*Total duration (days)*	*Start of work package (days)*	*Finish of work package (days)*
1 Strip formwork	6	120	0	120
2 Air conditioning ducts	8	160	12	172
3 Hydraulics	5	100	31	131
4 Internal brickwork	7	140	45	185
5 Concrete façade	5	100	96	196
6 Suspended ceiling	6	120	107	227
7 Interior finishes	8	160	119	279

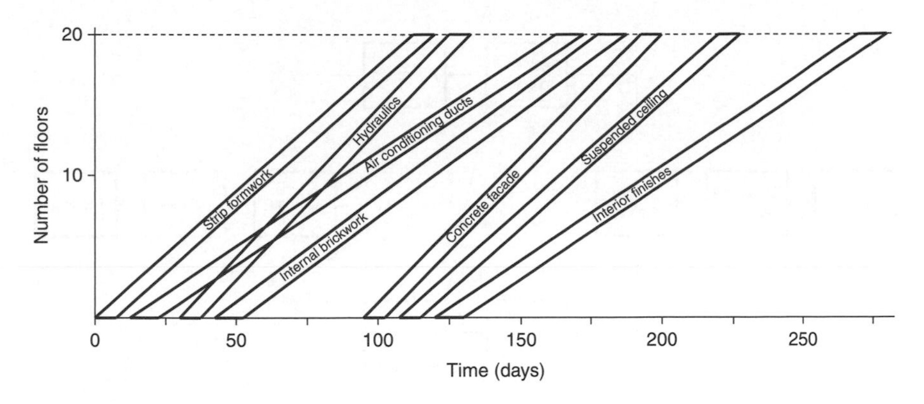

Figure 10.7 The LOB schedule of the fitout project.

package 'Interior finishes' to six days, the project duration would be shortened by around 20 days, and if all the packages had the same cycle time of, say, six days, approximately 60 days would be saved.

10.5 Developing a LOB schedule for projects requiring multiple crews

Tight construction schedules often require the planner to allocate more than one crew of workers to undertake a repetitive activity in order to speed up production. The following example will demonstrate a process of developing a LOB schedule with multiple crews.

A large housing project comprises 100 identical houses. The project must be completed within 60 weeks and the contractor is required to deliver three houses per week from week 25 onward. A precedence schedule showing the construction sequence of a typical house is given in Figure 10.8. The contractor, an experienced house builder, has an extensive database with up-to-date information on cost and time estimates of similar houses. From the data, the contractor has calculated the volume of person-hours required for accomplishing each activity. This information is given in Table 10.3.

Person-hours represent productivity rates of activities. They need to be converted to activity durations, crew sizes and actual rates of output. First, the contractor needs to estimate ideal sizes of crews necessary to perform the project activities. These are given in Table 10.4 in column 3. Next, a theoretical crew size necessary to perform the volume of work is calculated using the following formula:

$$\text{Theoretical crew size} = \frac{\text{Handover rate} \times \text{Person-hours per activity}}{\text{Number of working hours per week}}$$

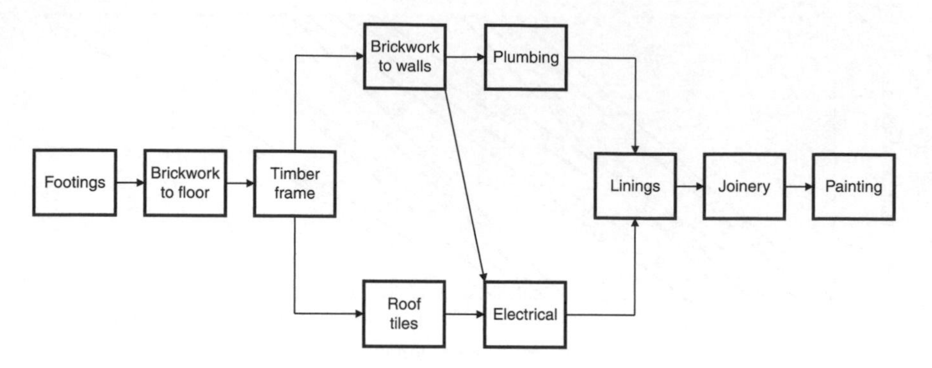

Figure 10.8 A construction schedule of a house.

Table 10.3 The volume of work per activity expressed in person-hours

Activities	*Person-hours per activity*
1 Footings	48
2 Brickwork to floor	120
3 Timber frame	160
4 Brickwork to walls	420
5 Roof tiles	96
6 Electrical	96
7 Plumbing	80
8 Linings	96
9 Joinery	128
10 Paint	96

Table 10.4 Crew sizes and the rate of output

Activities	*Person-hours per activity*	*Estimates of ideal crew size*	*Theoretical crew size*	*Actual crew size*	*Actual rate of progress*
Footings	48	2	3.0	4	4.0
Brickwork to floor	120	5	7.5	10	4.0
Timber frame	160	4	10.0	12	3.6
Brickwork to walls	420	5	26.3	30	3.4
Roof tiles	96	4	6.0	8	4.0
Electrical	96	2	6.0	6	3.0
Plumbing	80	2	5.0	6	3.6
Linings	96	4	6.0	8	4.0
Joinery	128	2	8.0	8	3.0
Paint	96	2	6.0	6	3.0

For the purpose of this example, let's assume that the project works six days per week, eight hours per day. The theoretical size of the 'Footings' crew is then $(3 \times 48)/(8 \times 6) = 3.0$. The theoretical crew sizes for the other activities are calculated in the same manner and are given in Table 10.4 in column 4.

Actual crew sizes are determined next by rounding the theoretical crew size to a number that is a multiple of the estimated ideal crew size. For example, the theoretical crew size of 'Footings', calculated as 3, is be rounded to 4, which is the multiple of the estimated crew size.

The next factor to determine is the actual rate of progress per week. It is calculated using the following formula:

$$\text{Actual rate of progress} = \frac{\text{Actual crew size} \times \text{Required handover rate}}{\text{Theoretical crew size}}$$

The actual rate of progress of the activity 'Roof tiles' is: $(8 \times 3)/6 = 4.0$. The actual rates of progress of the other activities have been calculated and are given in Table 10.4 in column 5.

The number of crews that need to be allocated to each activity can now be calculated from the following formula:

$$\text{Number of crews} = \frac{\text{Actual crew size}}{\text{Ideal crew size}}$$

For example, the number of crews to perform activity 'Joinery' is $8/2 = 4$. The calculated numbers of crews are given in Table 10.5 in column 2.

The next factor to be determined is duration per crew per activity (in days). It is calculated using the following formula:

$$\text{Duration per activity} = \frac{\text{Person-hours per activity}}{\text{Estimated crew size} \times \text{Number of working hours per working day}}$$

For example, duration of the activity 'Electrical' is: $96/(2 \times 8) = 6$ days. The remaining durations have been calculated and are given in Table 10.5 in column 3.

Finally, the total time required to complete a sequence of repetitive activities for the whole project is calculated from the following formula:

$$\text{Total activity duration} = \frac{\text{Number of houses} \times \text{Number of working days per week}}{\text{Actual rate of progress}}$$

The total duration of the activity 'Brickwork to floor' is: $(100 \times 6)/4.0 = 150$ days. The remaining total activity durations have been calculated and are given in Table 10.5 in column 4.

It is now possible to determine start and finish dates of the house activities

Table 10.5 Number of crews, duration of activities, total duration, and start and finish dates

Activities	*Number of crews*	*Duration per crew (days)*	*Total duration (days)*	*Start of activity*	*Finish of activity*
Footings	2	3	150	0	150
Brickwork to floor	2	3	150	3	153
Timber frame	3	5	167	6	173
Brickwork to walls	6	10.5	177	11	188
Roof tiles	2	3	150	26	176
Electrical	3	6	200	29	229
Plumbing	3	5	167	26	193
Linings	2	3	150	82	232
Joinery	4	8	200	85	285
Paint	3	6	200	91	291

from the information in Table 10.5, using the same approach as that defined in section 10.4 above. The total project duration is 291 days, or 49 weeks, and the first house will be completed in 96 days, or just over 16 weeks. The LOB tabular schedule in Table 10.5 shows that the calculated number of crews is about right to meet the contract delivery requirements.

However, before proceeding any further, it is important to examine the impact on the schedule by the uneven number of multiple crews that have been allocated to the activities. The activity 'Timber frame' will be performed by three crews of workers, while the preceding activity, 'Brickwork to floor', will be performed by two crews only. It means that on day 6, when 'Timber frame' crews are to begin work, only two 'Brickwork to floor' activities will be completed. In order to start all three 'Timber frame' crews, it is necessary to delay the start of this activity until day 11 when enough 'Brickwork to floor' activities have been completed. The finish date of the activity 'Timber frame' will now be day 178.

Similarly, 'Brickwork to walls' employs six crews of workers while the preceding activity, 'Timber frame', employs only three. Three 'Timber to walls' activities will be completed on day 16 and six on day 21. This means that the activity 'Brickwork to walls' will now start on day 21 with all six crews and will be completed on day 198.

The start dates of the activities 'Electrical' and 'Joinery' will also need to be adjusted to match the rate of progress of their preceding activities. The adjusted start and finish dates of the activities are given in Table 10.6.

The project will now be completed in 302 days, or 51 weeks, with the first house delivered on day 108, or week 18.

Finally, let's add six-day buffer zones to the schedule for unforeseen delays, and recalculate start and finish dates of the activities. The result is given in Table 10.7.

Table 10.6 Start and finish dates adjusted for multiple crews

Activities	*Number of crews*	*Duration per crew (days)*	*Total duration (days)*	*Start of activity*	*Finish of activity*
Footings	2	3	150	0	150
Brickwork to floor	2	3	150	3	153
Timber frame	3	5	167	11	178
Brickwork to walls	6	10.5	177	21	198
Roof tiles	2	3	150	31	181
Electrical	3	6	200	37	237
Plumbing	3	5	167	36	203
Linings	2	3	150	90	240
Joinery	4	8	200	96	296
Paint	3	6	200	102	302

Table 10.7 The final start and finish dates with buffer zones

Activities	*Total duration (days)*	*Start of activity*	*Finish of activity*
Footings	150	0	150
Brickwork to floor	150	9	159
Timber frame	167	17	184
Brickwork to walls	177	27	204
Roof tiles	150	43	193
Electrical	200	55	255
Plumbing	167	48	215
Linings	150	114	264
Joinery	200	126	326
Paint	200	138	338

The contractor can be reasonably confident of completing the project in 338 days, or 57 weeks, and delivering the first batch of houses on day 144, or week 24. The contractor's schedule meets the contract requirements of completing 100 houses in 60 weeks with the delivery of three houses per week commencing from week 25 onwards. The LOB schedule for the project is shown graphically in Figure 10.9.

10.6 Summary

This chapter described the LOB scheduling method. The main benefit of LOB lies in scheduling repetitive projects. It is particularly useful in determining delivery dates of repetitive activities of projects. LOB is a resource-based

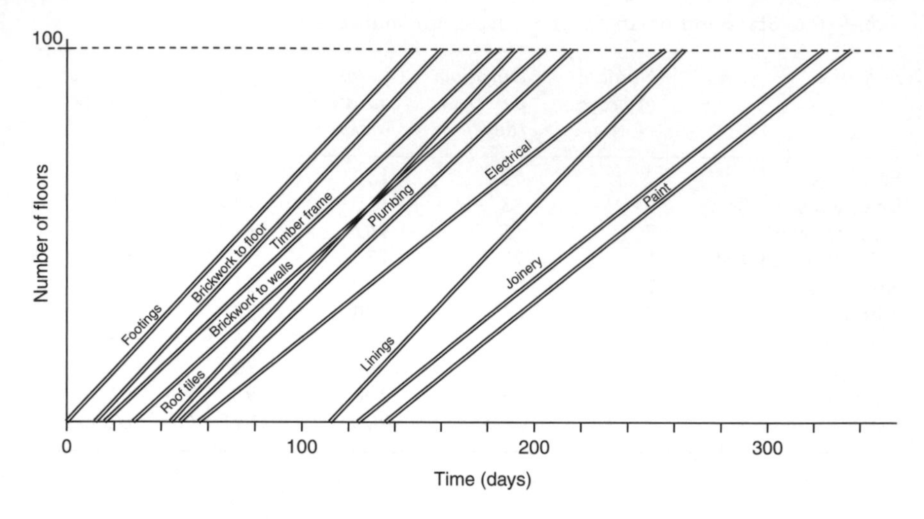

Figure 10.9 The LOB schedule for the housing project.

scheduling method that ensures that resources allocated to repetitive activities are employed efficiently.

Exercises

Solutions to the following exercises can be found on the following website: http://www.routledge.com/books/details/9780415601696/

Exercise 10.1

Prepare a LOB schedule (table and graph) for the construction of 40 typical floors of a high-rise commercial building. A precedence schedule showing the sequence of a typical floor construction is given in Figure 10.10. Each activity will be performed by one crew of subcontractors. Durations are in days. Add time buffer zones of six days between the activities.

Exercise 10.2

A contractor has been awarded a contract to refurbish 100 service stations and deliver them to the client at the rate of two per week after week 25. The project must be completed within 75 weeks.

Prepare a LOB schedule (table and graph) for this project. Assume a six-day working week. A precedence schedule showing the sequence of one service station construction is given in Figure 10.11. Durations are in weeks. Add time buffer zones of one week between the activities.

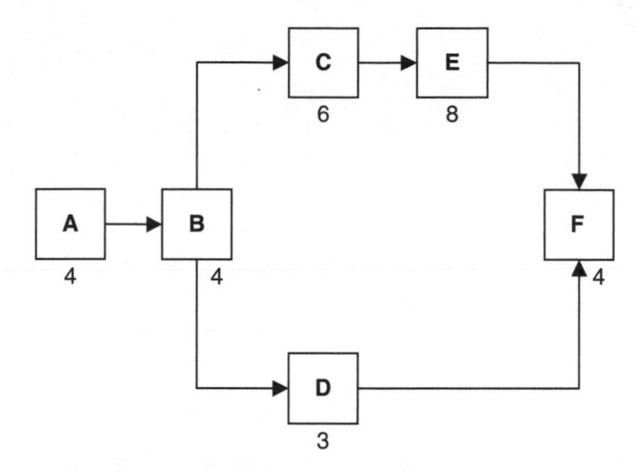

Figure 10.10 The precedence schedule for the example.

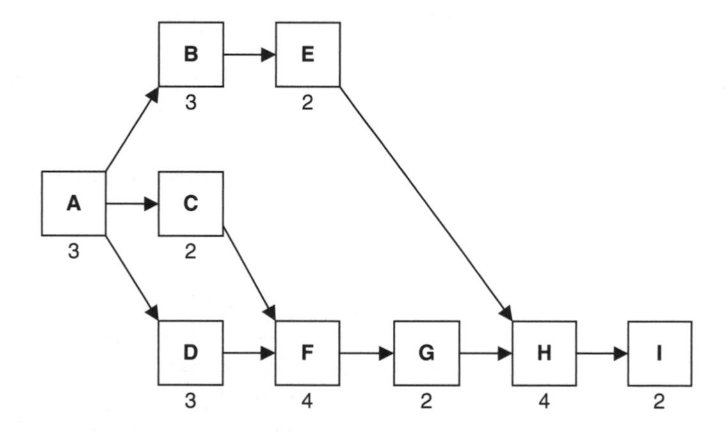

Figure 10.11 The precedence schedule for the example.

Chapter 11

Work study

11.1 Introduction

The purpose of this chapter is to develop an understanding of the concept of work study, which is concerned with improving the productivity of production processes. Its two principal components, method study and work measurement, will be examined in detail and will be supported by practical examples.

The previous chapters of this book have examined various scheduling techniques suitable for use in the construction industry. They are generally regarded as 'operations research techniques'. 'Operations research' focuses on applying scientific methods and tools to problems involving the operation of a system and developing optimum solutions. It is a category of 'operations management', which is concerned with planning, organising and controlling business operations, systems and models.

'Work study' is also a category of 'operations management' but is fundamentally different from 'operations research'. Work study aims at improving productivity by examining in detail specific parts of a system rather than the system as a whole. It increases productivity through better work methods, an improved use of resources and better management practices. Although not a scheduling technique, work study nevertheless requires the application of scheduling skills in developing more productive production processes.

In improving productivity, work study helps in identifying weaknesses in current work methods and in developing appropriate corrective strategies on which the best overall work method is then based. Work study helps the manager to examine the degree of utilisation of committed resources such as people, plant and materials. It also involves examination of management practices, such as empowerment, communication, planning efforts, motivation of personnel, and incentive schemes, on which production processes depend.

'Productivity' is at the heart of work study. It is defined as the ratio between output and input:

$$\text{Productivity} = \frac{\text{Output}}{\text{Input}}$$

'Output' is a rate of production or the work produced, while 'input' refers to the combination of resources, work methods and management practices necessary to make the production process work. Improved productivity can be achieved either by increasing 'output' while trying to keep 'input' constant, or by reducing 'input' while trying to keep 'output' constant. The maximum improvement in productivity could ideally be achieved by increasing 'output' while at the same time reducing the volume of 'input'. In this chapter, the emphasis will be placed on achieving improvement in productivity through better utilisation of labour and plant resources, and improved work methods. Given that most building projects are located on confined and often poorly accessible sites, and are required to accommodate a varying number of workers and wide range of construction plant, knowledge of the key principles of work study among the members of a project team is vitally important in formulating the most effective and efficient work methods, and in managing the human and physical resources.

The labour resource is by far the most difficult resource to manage, particularly in the construction industry, where people from different companies, different ethnic backgrounds, different professions, and different levels of experience and skills come together to form a short-term project team. People's performance rates tend to vary because no two persons are alike in experience, motivation and skill. People may also strike, take time off, get sick or simply not be available when required. Forcing people to work harder and faster is unlikely to result in a sustained increase in the production output unless people are highly motivated, appropriately skilled and adequately rewarded, and unless most-appropriate work methods have been installed.

The emphasis in managing construction plant/equipment is on selecting the most appropriate type and maintaining its full utilisation. Because technical and performance specifications of plant/equipment are known for a wide range of tasks and applications, estimating and maintaining their production rates should not be a problem provided the right plant/equipment is selected and installed as part of an efficient work method.

Work study embraces the parallel techniques of 'method study' and 'work measurement' (Figure 11.1). These will now be discussed in more detail.

11.2 Method study

The aim of method study is to analyse work methods, systems and procedures currently in use in order to develop improvements. There is always more than one method of work that could be employed. The role of a project manager is to examine as many work methods as possible in order to develop an easier and more effective production process. The project manager is required to:

- Have knowledge of the construction site and its limitations in order to prepare an effective site plan showing the layout and the design of the workplace
- Understand how the construction work is to be carried out and with what resources

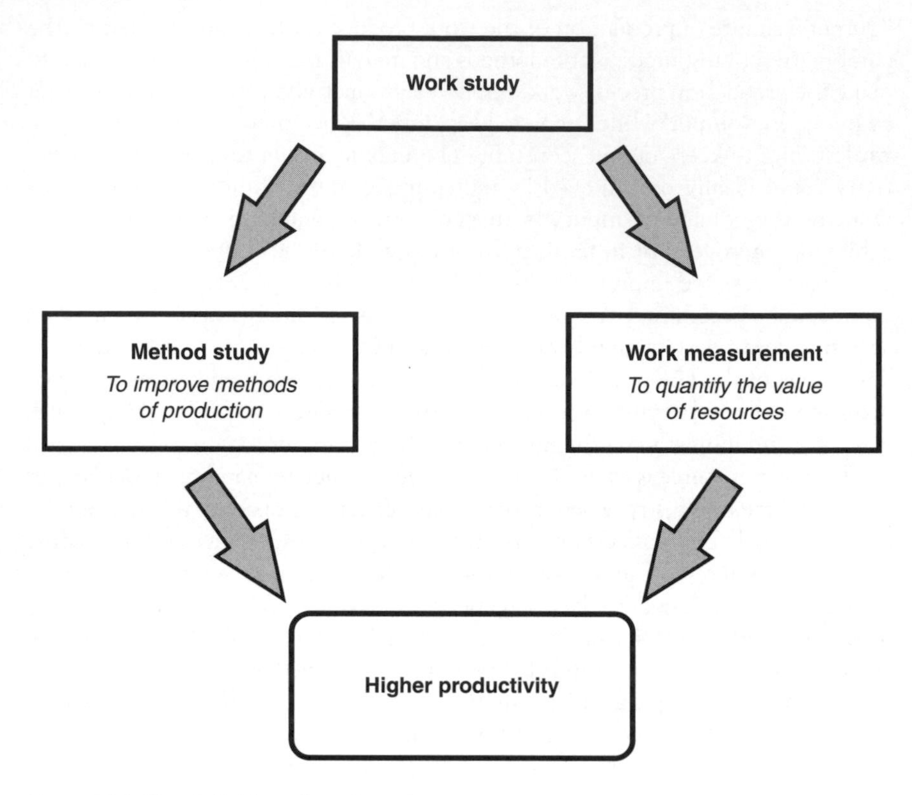

Figure 11.1 The definition of work study.

- Ensure maximum utilisation of resources
- Minimise delays and waste
- Be familiar with the design and the specification of the end product.

The general procedure for performing method study involves a number of discrete steps including:

- Defining the problem and selecting the work to be studied
- Recording the observed facts
- Examining these facts critically and seeking alternative solutions
- Developing the most practical, economical and efficient work method
- Installing the improved work method as standard practice
- Maintaining the improved method.

The individual steps in the method study procedure will now be examined in more detail.

11.2.1 Defining the problem

Before a problem can be solved, it must first be identified. This is rarely as straightforward as it might appear. For example, a long delay in unloading trucks outside a construction site may be seen as being caused by the lack of capacity of the site crane. However, poor scheduling of the work by the planner may be the actual main cause. Committing an additional crane would probably solve the problem in the short term, but the problem would most likely reappear in the future and undue costs may be added to the project. The underlying cause of the problem must ultimately be fixed.

Experience will guide the project manager to examine and monitor those aspects of the work that are likely to have the greatest impact on productivity. By systematically sorting through a range of technical, economic (cost and time) and human factors, the project manager will gain a better understanding of the problem and its underlying causes.

11.2.2 Recording the facts

A systematic examination of the identified problem and its underlying causes generates valuable information not just on the problem itself and its causes but also on the adopted method of work. This information assists the project manager in assessing the viability of alternative solutions. If, for example, the identified problem is the lack of capacity of personnel hoists in moving workers vertically throughout the project, before the project manager embarks on implementing a solution to fix this problem, the project manager would investigate the underlying causes of the problem. They may be related, for example, to inadequate scheduling of work, the larger than expected size of the workforce, or poor serviceability and/or location of the hoists. The information thus generated is invaluable for developing a solution to the problem. It is essential that it is recorded in the most appropriate manner. For this purpose, work study adopts a variety of charts and diagrams such as process charts, flow charts and multiple activity charts. Let's examine these charts in detail.

Process chart

A process chart is useful for representing a sequence of activities in a production process using standard symbols (see Figure 11.2). For example, a process of placing concrete is graphically represented by the process chart symbols in Figure 11.3.

This format of data recording allows the project manager to aggregate activities into their respective groups for further analysis. Productive activities are 'operations'. 'Transport' activities are generally regarded as unproductive even if they form an essential part of the production process. The activity 'concrete kibble lifted by crane' in Figure 11.3 is a transport activity, which is unproductive. The remaining activities are also regarded as unproductive.

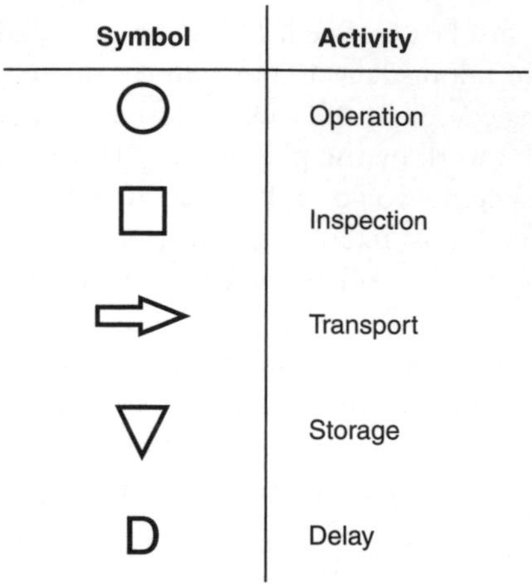

Symbol	Activity
○	Operation
□	Inspection
⇨	Transport
▽	Storage
D	Delay

Figure 11.2 Standard symbols used in process charts.

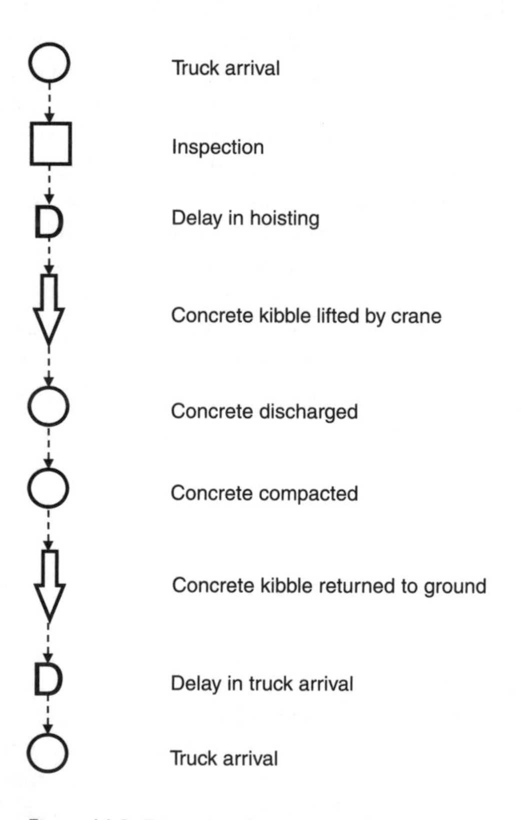

Figure 11.3 Example of a process chart.

Apart from determining a ratio between productive and unproductive activities, the project manager also gains better insight into the actual production process and its components that assist the project manager in identifying potential weaknesses and problems. The subject method of placing concrete has three productive and five unproductive activities. Clearly, the presence of two delay activities in Figure 11.3 highlights the specific weakness of the method.

Flow process chart

A flow process chart records more complex information about current and alternative work methods. It traces the flow of work in the process by connecting corresponding standard process chart symbols. It also:

- Aggregates activities within the production process into standard symbol groups
- Measures the production process in specific units such as quantity of materials, distances travelled or activity times
- Summarises the present and alternative methods and their costs.

An example of a flow process chart is given in Figure 11.4, which shows a method of fabricating timber roof trusses. The chart is divided into three distinct parts:

- The top left corner defines the process, its location and the responsible party
- The top right corner summarises information generated by the flow process chart
- The bottom part defines the logic of the production process, measures it in appropriate units, and aggregates it according to the standard symbol groups. It also shows the flow of work.

Let's look at the bottom part of the flow process chart in Figure 11.4 first. A brief glance at the total of individual aggregated activities suggests that out of 14 activities, only three are productive. Clearly, the present production method seems inefficient, suffering from discontinuity of work, delays and frequent movement of materials. It also shows that the total length of the production process is 95 metres. An improved production method is given in Figure 11.5. The delays have been eliminated and the number of transportations and storage activities reduced. The overall length of the production process has also been reduced by 15 metres. The top right corner of the chart provides a summary of the analysis.

Multiple activity chart

Multiple activity charts (MACs) have already been described in Chapter 9. Their ability to visually assess the use of committed resources makes them highly suitable for recording and analysing method study information.

Flow process chart

	Summary			
Chart no.: 5 Sheet no. : 2	Activity	Present	Alternative	Saving
Product: House construction	○ □ ⇨			
Activity: Fabrication of timber trusses Method: Present	▽ D			
	Distance (m)			
Location: 25 White St., Kingsford 2032 NSW	Time			
Operator: J. Brown	Cost of: labour material			
Charted by: H. Smith Approved by: M. Strong	Total $			

Activities	Quantity	Distance (m)	Time (min.)	Symbols ○	□	⇨	▽	D	Remarks
Timber delivered						●			unproductive
Unloaded				●					productive
Taken to store		30				●			unproductive
In storage							●		unproductive
To cutting station		10				●			unproductive
Awaiting cutting								●	unproductive
Cut to length				●					productive
To fabrication station		30				●			unproductive
Awaiting fabrication								●	unproductive
Fabricate trusses				●					productive
Awaiting inspection								●	unproductive
Inspection					●				unproductive
Taken to store		25				●			unproductive
In storage							●		unproductive
Total		95		3	1	5	2	3	

Figure 11.4 Example of a flow process chart: the current method of work.

Flow process chart

Chart no.: 5 Sheet no. : 2	Summary			
	Activity	Present	Alternative	Saving
Product: House construction	○	3	3	0
	□	1	1	0
	⇨	5	4	1
Activity: Fabrication of timber trusses	▽	2	1	1
Method: Improved	D	3	0	3
	Distance (m)	95	80	15
Location: 25 White St., Kingsford 2032 NSW	Time			
Operator: J. Brown	Cost: labour material			
Charted by: H. Smith Approved by: M. Strong	Total $			

Activities	Quantity	Distance (m)	Time (min.)	○	□	⇨	▽	D	Remarks
Timber delivered						●			unproductive
Unloaded				●					productive
To cutting station		25				●			unproductive
Cut to length				●					productive
To fabrication station		30				●			unproductive
Fabricate trusses				●					productive
Inspection					●				unproductive
Taken to store		25				●			unproductive
In storage							●		unproductive
Total		80		3	1	4	1	0	

Figure 11.5 Example of a flow process chart: the improved method of work.

Flow chart

A flow chart is another tool that helps to visually record information pertinent to method study. It is most effective in displaying the flow of work within a workplace, such as construction sites, design offices and factory production floors. A line joining the standard process chart symbols that define the location of specific tasks within the production process represents the flow of work. A flow chart in Figure 11.6 shows the delivery, storage and distribution of bricks on a construction

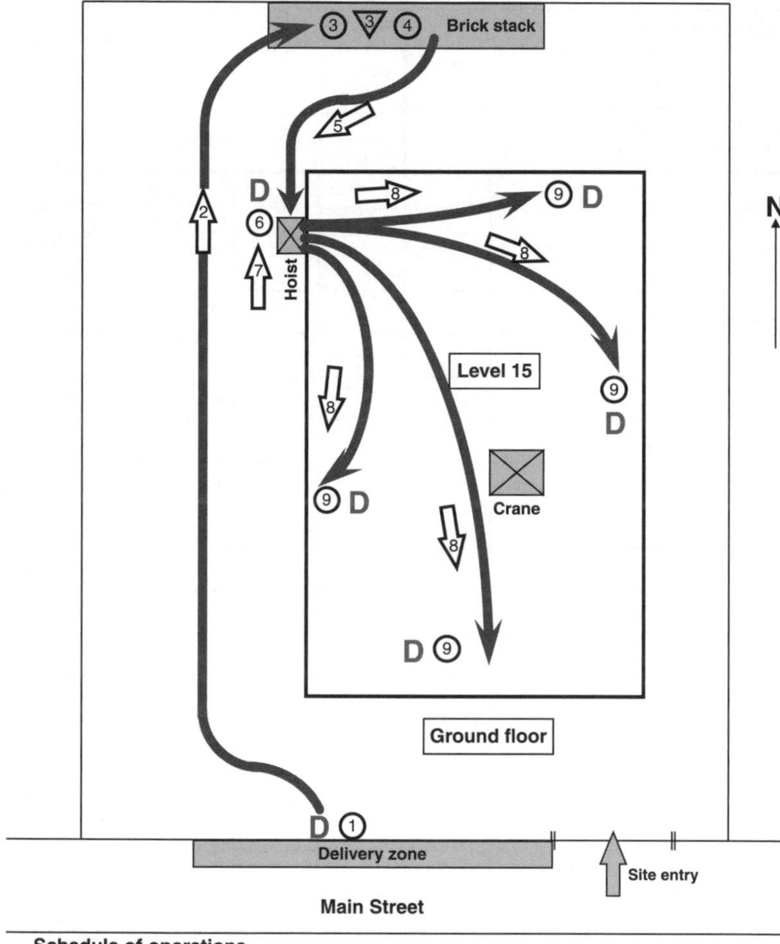

Figure 11.6 Example of a flow chart.

site. There may be valid reasons for the present location of the hoist and the storage area for the bricks. But it would certainly be worth investigating whether both the hoist and the brick storage area could be relocated closer to the southern end of the site to reduce the amount of movement of bricks.

11.2.3 Examining facts and seeking possible alternative solutions

The examination stage is the most important part of method study and requires the project manager to take on a challenging attitude and an impartial judgement in order to examine systematically and objectively the recorded information concerning the work method under investigation. The project manager makes a detailed and critical review of each activity with the aim of developing a more efficient work method by eliminating any unnecessary tasks, preventing delays, reducing movements of resources and simplifying work tasks.

In developing a new work method, the project manager is expected to:

- Examine the facts
- Discard preconceived ideas
- Display challenging attitude
- Avoid hasty judgements
- Look for details
- Discard undesirable elements of the present work method.

11.2.4 Developing the most practical, economical and efficient method

In this stage of method study, the project manager selects the most appropriate work method from the information generated in the previous stages. In doing so, the project manager needs to satisfy the following principles:

- Employ most appropriate resources, both human and physical
- Maximise utilisation of committed resources
- Minimise the extent of human movement
- Minimise the extent of movement of materials and plant/equipment
- Maintain environmental and safety standards
- Develop the most efficient, practical and safe layout of the workplace
- Implement safe working procedures.

11.2.5 Installing the improved method

At this stage the project manager has selected a preferred modified or alternative method and is ready to install it. A minor modification to the present method of work may often be implemented swiftly and with little or no disruption to the production process. But implementation of a more substantially modified method or an alternative method requires careful planning to ensure smooth changeover with minimum disruption to the production process. It should be carried out in two phases: preparation and installation.

Preparation

In this phase, the project manager develops a plan of action for modifying or replacing the current work method. It is essential that those affected by the proposed change are involved in the development of such a plan. It is also essential that the reasons for the change and the details of the proposed plan of action are communicated throughout the workplace. People are generally inclined to reject change, particularly if they don't know the reasons for change and what impact it will have on them. Enforcement of a change in the workplace by the management with little or no involvement of the workers may in extreme cases lead to industrial unrest.

In developing a plan of action, the project manager would need to consider a range of issues including:

- Timetable of the work
- Availability of resources
- Replacement of the old stock
- Training of labour
- Compliance with safety regulations and labour awards.

Where possible, a major change to the work method should be carried out outside normal working hours. In some cases, however, it may be necessary to close down the entire production process for a time. For example, introduction of a new car model requires a complete shutdown of the car maker's production line for a number of weeks to carry out retooling.

In situations where disruption to the production process must be kept to the absolute minimum, rehearsal of a changeover may need to be seriously considered first.

Installation

With a good plan in hand, the installation phase of a changeover should become a routine task. Monitoring the performance of a new work method is an important aspect of installation. Monitoring seeks to uncover:

- Any deviations in the performance of a new method, which may require fine-tuning the method or more significant modifications
- An unexpected depletion of the resources
- Changes in workers' attitudes and motivation.

11.2.6 Maintaining the improved method

It is the responsibility of the project manager to regularly review the performance of a new work method and attend to minor or major corrections of the method as required.

11.3 Work measurement

'The aim of work measurement is to determine the time it takes for a qualified worker to carry out a specific job at a defined level of performance and to eliminate ineffective elements of work' (Oxley and Poskitt 1980: 161). The principal aim is to seek the standard time of work, which can assist in (Harris and McCaffer 2006):

1 Determining appropriate quantities of human and physical resources; once standard times of activities in a construction schedule are known, it is then possible to allocate appropriate resources to such activities (see Chapter 4)
2 Measuring the utilisation of committed resources
3 Providing the basis for sound financial incentive schemes
4 Evaluating the economic viability of alternative methods of work.

The latter three issues listed above will be discussed briefly later in this chapter. Let's focus first on determining the standard time of work.

11.3.1 The standard time of work

There are several well-known techniques used for work measurement: time study, time and motion study, activity sampling and synthetical estimating (Currie 1959; Oxley and Poskitt 1980; Harris and McCaffer 2006). Since time study is recognised as the basic measuring technique, it will be discussed in more detail.

The technique of time study determines the standard time of work, which is the time that an average qualified worker who is sufficiently motivated, instructed and supervised would take to complete the task. While the standard time of work is commonly derived from databases, it is nevertheless useful to examine a method of deriving it from a limited number of observations by a stopwatch.

The standard time of work can be calculated from the basic time, relaxation allowances and contingencies using the following formula:

$$\text{Standard time} = \text{Basic time} + \text{Relaxation allowances} \\ + \text{Contingency allowance}$$

Basic time

'Basic time' is the observed time measured by a stopwatch and rated for a defined standard of performance. If the manager is able to measure the time of a specific activity performed by a large sample of workers, the project manager will be able to calculate an arithmetic mean of all the measurements of the sample and use it as basic time without any adjustments. But this approach would be time-consuming and costly. A more practical approach is to measure the performance of a single worker, though in this case the worker under study may or may not represent the 'norm' or the 'standard level of performance'. The project manager would need to make a subjective judgement about the observed level of performance and relate it to its own perception of what constitutes a normal or standard level of performance. This process is referred to as 'rating'.

'Standard performance is the working rate of an average qualified worker who is sufficiently motivated, instructed and supervised' (Currie 1959: 38). This worker is able to maintain the speed of work day after day without undue physical or mental fatigue. A rating of 100 is commonly assigned to such standard performance (BS 1992). Ratings above 100 are associated with a worker who works faster, while ratings below 100 refer to one who works slower. For example, if 100 is the standard rating, then:

- 125 relates to a worker who is very quick, highly skilled and highly motivated
- 75 relates to a worker who is not fast, with average skill and uninterested
- 50 relates to a worker who is very slow, unskilled and unmotivated (Harris and McCaffer 2006).

It is now possible to calculate basic time from the following formula:

$$\text{Basic time} = \text{Observed time} \times \frac{\text{Assessed rating}}{\text{Standard rating}}$$

Assume that the project manager timed a specific task performed by a worker to be ten minutes. This is the observed time. The manager assessed the performance of the worker as 125. The basic time of that specific task performed by the worker is then:

$$\text{Basic time} = 10 \text{ minutes} \times (125/100) = 12.5 \text{ minutes}$$

Because the project manager assessed the worker's performance as better than the norm, an average worker would take more time to perform this task. Hence basic time is 12.5 minutes. If the project manager's assessed rating of the worker was, say, 80, the basic time would then be:

$$\text{Basic time} = 10 \text{ minutes} \times (80/100) = 8 \text{ minutes}$$

In this case, the worker under observation is slow and an average worker would therefore be expected to accomplish the task faster.

Relaxation allowances

'Relaxation allowances' provide for the time lost due to a wide range of causes such as:

- The worker attending to personal needs
- The worker relaxing, resting or recovering from the effects of fatigue
- The worker slowing down due to external influences such as weather, noise, light or other conditions.

Determination of the magnitude of relaxation allowances relevant to construction activities has not yet been adequately addressed. When used, they are expressed as a percentage of the basic time. Example relaxation allowances can be found in Harris and McCaffer (2006).

Contingency allowance

'A contingency allowance' provides for the time lost due to other factors such as maintenance of plant/equipment, waiting time, breakdowns, unexpected site conditions, variations, etc. (Harris and McCaffer 2006). The amount of contingency is expressed as a percentage of basic time.

Calculation of standard time of work

With all the components of the standard time equation now defined, let's calculate the standard time of work from the following information:

Basic time is 20 minutes

Relaxation allowances are 20% of basic time = $(20 \times 20)/100 = 4$ minutes

Contingency allowance is 15% of basic time = $(15 \times 20)/100 = 3$ minutes

The standard time of work is then:

Standard time = 20 min. + 4 min. + 3 min. = 27 minutes

11.3.2 Measuring utilisation of committed resources

With the knowledge of the standard time of work, it is possible to measure utilisation of committed resources, particularly plant/equipment and labour. The following formula defines utilisation:

$$\text{Utilisation} = \frac{\text{Total standard time of work}}{\text{Time of work available}} \times 100\%$$

Let's demonstrate the application of the above formula to a simple practical example. A project in question is a three-storey residential building. A single barrow hoist has been installed to move all the required materials to each level of the building. What is the present utilisation of the hoist based on the following information?

- Basic time for one lifting cycle of work of the hoist is five minutes (it means that it takes five minutes to move the hoist's platform up and return it to the original location)
- Relaxation allowances are 25% of basic time = (25 × 5)/100 = 1.25 minutes
- Contingency allowance is 20% of basic time = (20 × 5)/100 = 1.0 minute
- At the end of day 3 the hoist had made a total of 115 lifting cycles (based on an eight-hour working day).

$$\text{Standard time} = 5 \text{ min.} + 1.25 \text{ min.} + 1.0 \text{ min.} = 7.25 \text{ minutes}$$

$$\text{Utilisation} = (7.25 \text{ min.} \times 115 \text{ lift cycles})/(3 \text{ days} \times 8 \text{ hours} \times 60 \text{ min.}) \times 100\% = 57.9.$$

The hoist's utilisation is presently 57.9%.

11.3.3 Providing the basis for sound financial incentive schemes

The knowledge of the standard time of work makes it possible to determine the volume of work to be accomplished within a given period. Specific work targets may then be set defining what the workers would need to accomplish in that given period. To ensure that these work targets are met and that a high level of productivity is maintained, it may be appropriate to offer financial incentives to the workers. The information on financial incentives can be found in most management textbooks, including Pfeffer (1998).

11.4 Activity duration

One of the most difficult tasks in planning is establishing the duration of activities or the standard time of activities. In section 11.3 the concept of time study and the standard time was reviewed. It would, however, be impractical and time-consuming to base estimation of activity duration on that concept. It is expected that organisations working in the construction industry have access to a database of productivity rates and output rates of plant and equipment from which standard time of work in the form of activity duration are calculated. Such calculations are then based on the quantity of the work to be carried out, the resources needed for its execution, the specific contractual requirements imposed on the project and the presence of risk.

11.4.1 Determining time duration of activities from labour productivity rates

The quantity of work is commonly measured and compiled by a quantity surveyor in a document called a 'bill of quantities'. In some countries, such as the USA and Japan, clients do not commission consulting quantity surveyors to prepare a bill of quantities as a bidding document; rather, bidding contractors are required to prepare their own quantities.

When a bill of quantities for a particular project is available to the contractor, the contractor's planner can easily determine the volume of work for each activity in the project. For example, the bill of quantities specifies the quantity of 'trench excavation' as 60 m^3. Let's assume that the trench is 15 metres long, 2 metres wide and 2 metres deep, and the soil is clay.

If this excavation activity is to be performed by labourers, the planner needs to know a productivity rate or units of work for labour excavating the trench. The planner may deduce this productivity rate from experience or extract it from databases. The task of calculating duration of this activity is fairly simple once the planner has determined the total volume of labour hours and the size of the labour crew:

Total labour hours = quantity of work × productivity rate

Activity duration = total labour hours/number of persons

Assume that 1.5 labour hours is required to excavate 1 m^3 of soil. Therefore,

Total labour hours = 60 m^3 × 1.5 labour hours

= 90 labour hours or 12 labour days (at 8 hours per day)

With, say, three persons assigned to this activity, duration will be:

Activity duration = 12/3 = 4 days

The planner may vary the activity duration by either increasing or decreasing the labour crew size, provided this is possible or practicable. For example, with two and four people assigned to the above activity, its duration would be six and three days respectively. The planner may then optimise alternative activity durations in terms of cost and time to determine the optimum outcome. Cost–time optimisation is examined in Chapter 6.

11.4.2 Determining time duration of activities from daily output rates of resources

Duration of activities can also be calculated from the volume of work and the daily output rates of resources. In most countries, output rates per day for plant/

equipment and labour may be obtained from published cost data catalogues. The calculation process to determine activity duration in days is:

Activity duration in days = quantity of work/resource output rate per day

For the previous example, assume that a backhoe will be used to excavate the trench. The output rate of the selected backhoe is, say, 80 m^3 per day. To excavate the required 60 m^3 will be calculated as follows:

Activity duration = 60/80 = 0.75 day

Although it takes less than one day to excavate the trench, the planner will probably round the duration to the nearest day, in this case one day. More complex tasks performed by plant or equipment may require the allocation of additional labour to assist with and supervise such tasks.

11.4.3 Labour productivity and daily output databases

When estimating the duration of activities, the use of accurate 'labour productivity 'and 'daily output' rates (rates) is fundamental to producing reliable time schedules. The rates contained in databases that were not produced by the planner often lack details on the project particulars from which they were derived. The project particular information which affects rates includes information such as:

- Labour requirements
- Productivity/efficiency of the resources carrying out the work
- Project geographic location
- Site access conditions
- Work environment
- Changes in technology.

When using databases that were not produced by the planner, the uncertainty associated with the project particulars makes it difficult for the planner to determine whether the rates in the database are applicable to the project being estimated. Due to this difficulty, it is not uncommon for planners to build their own databases by recording data from actual completed past projects that they experienced and are familiar with. As rates are primarily derived from 'labour requirements' – which do not vary greatly over time – published and reputable databases can generally be relied upon when estimating the duration of activities.

When developing rates databases, the planner needs to ensure that the recording of project data is clear and accurate. The type of information that needs to be recorded from the start of the project includes:

- 'Work Method Statements' that were used in calculating the original activity durations.
- Activities' actual duration, resources and concurrent activities related to completed activities.
- Any stoppages or delays experienced with the activities.
- Information particular to a project.
- Outcomes of consultations with organisations carrying out actual work
- 'Method Study' results.

The recorded information is then compared with the original Work Method Statements from which a rate is calculated. The rate would be categorised in the database by factors such as geographic location and trade, with activity and project particulars also noted. The planner may also define a range of activity durations as opposed to one single rate; this range would typically include three rates for an optimistic, most likely and pessimistic scenarios. It is important for the planner to continually review and update the databases.

There is not a wide range of published databases available. In Australia one of the leading databases is the 'Rawlinsons Process Engineering Handbook'. The Rawlinsons database consists of rates derived from actual data, i.e. installation times taken from schedules used to track the processes on real construction sites, and where no actual data were available on a similar item, a rationally calculated extrapolation of time from the known data.

There is a growing trend for project planners to use dedicated project planning website databases produced by the international planning community. Members of these websites will upload rates from personal project experiences. Planners need to exercise caution when using these databases as there is usually no review process guaranteeing their accuracy and reliability.

11.4.4 Determining duration of activities from the target dates

Sometimes a bidding contractor is required to prepare a construction schedule that meets a tight completion date. The contractor's planner would need to schedule the project by working from its completion date to the start date by fitting all the work within the given time-frame. This often results in some activities having unrealistically short durations. In such cases, the planner is required to allocate whatever resources are needed to meet the required activity duration. One way of identifying and quantifying potentially unrealistically short durations is to prepare an initial schedule disregarding the target dates and then perform a series of project compressions until the target dates are met. Provided the target dates can be met, the project planner is then able to quantify the costs associated with meeting the target dates. Refer to Chapter 6 for an examination of a schedule compression.

11.4.5 Single-value estimates of activity duration

Before leaving this topic, it is useful to note that productivity and output rates of resources from which activity durations are calculated are commonly expressed as single-value estimates. They are extracted from databases in the form of mean or average rates. For example, an average output rate for laying bricks extracted from a database may be 300 per bricklayer per day. This rate has been compiled from virtually thousands of past bricklaying output rates stored in the database. Given that the mean or average is 50 per cent, the output rate of 300 bricks per bricklayer per day has only 50 per cent probability of being achieved. Therefore, the probability of achieving the rate of 300 bricks per day per bricklayer appears to be rather small. Scheduling that relies on using average productivity and output rates of resources is referred to as single-value (deterministic) scheduling. As is evident from the above example, relying on average time estimates is clearly risky.

An alternative approach known as probability (stochastic) scheduling expresses estimates of productivity and output rates in the form of probability distributions. An average estimate of the output rate in the database is in fact a mean of a distribution of the rates from which it was compiled. The range of the output rate distribution is defined by its standard deviation. When the standard deviation is high, the range is correspondingly wide. For example, with the standard deviation of 50 bricks per day, there is approximately 68 per cent probability that the range of the output rate would be between 250 and 350 bricks per bricklayer per day (i.e. the range between +1 to −1 standard deviation from the mean) and almost 100 per cent probability that it would be in the range between 150 and 450 bricks per bricklayer per day (i.e. the range between +3 to −3 standard deviations from the mean). While the probability is high, this range is wide and the actual output rate could in fact be anywhere within the defined range. Assuming that the output rate distribution is normal, by applying simple statistics it emerges that there is only about 16 per cent probability that one bricklayer would lay at least 350 bricks per day or about 84 per cent probability that one bricklayer would lay no more than 250 bricks per day. Using an appropriate probability analysis method such as the Monte Carlo simulation, the combined effect of individual probability distributions of the activities in a schedule can be assessed statistically. A brief introduction to probability estimating and the technique of the Monte Carlo simulation are discussed in Chapter 12.

11.5 Evaluating the economic viability of alternative methods of work

The concepts of method study and work measurement will now be applied to two practical examples assessing the economic viability of alternative methods of work.

11.5.1 Example 1

This example assesses the present work method of lifting formwork by a tower crane on a high-rise commercial project. The crane has been assigned to this task for three hours per day. The project manager has calculated the volume of formwork required to be lifted in the allocated three-hour time-slot as 20 loads (assuming one load per lift). To meet the construction schedule, the work method must ensure that the crane can deliver 20 loads of formwork in three hours.

Two crews of workers have been assigned to this task. One crew is available on the level where formwork is being stripped. Its task is to attach loads of formwork to the crane. Another crew is available on the working deck with a task of unloading the crane.

A list of specific activities performed by the crews of workers, including the standard times of work, are given in Table 11.1. The cost of the committed resources is given in Table 11.2.

The present method of work is illustrated in Figure 11.7 using a multiple activity chart. It is based on using only one set of slings (one set comprises two separate slings). Is the present method sufficient to meet the project requirements?

The MAC schedule in Figure 11.7 shows that the work is performed at a regular cycle of 12 minutes. It means that the current method is able to lift one load of formwork every 12 minutes. However, in three hours the current method of work can only handle 15 loads, which is well short of the required target of 20 loads. It is also worth noting that neither the bottom nor the top crew of workers is effectively utilised. The bottom crew achieves only 40 per cent utilisation while the top crew is utilised 50 per cent of its time. Utilisation of the crane while engaged in lifting formwork is at the 80 per cent level.

Table 11.1 Example 1 activities and the standard times of work

Activities	*Standard time (minutes)*
Attach a set of slings to one load of formwork	2
Detach a set of slings from one load of formwork	2
Hook and unhook a set of slings to and from crane	1
Lift one load by crane	2
Return a set of slings by crane	1
Manoeuvre load (attached to crane) into position	1

Table 11.2 Example 1 costs of committed resources

Resources	*Cost/day*
Crane	$5,000
1 Crew of workers	$1,000
1 Set of slings	$100

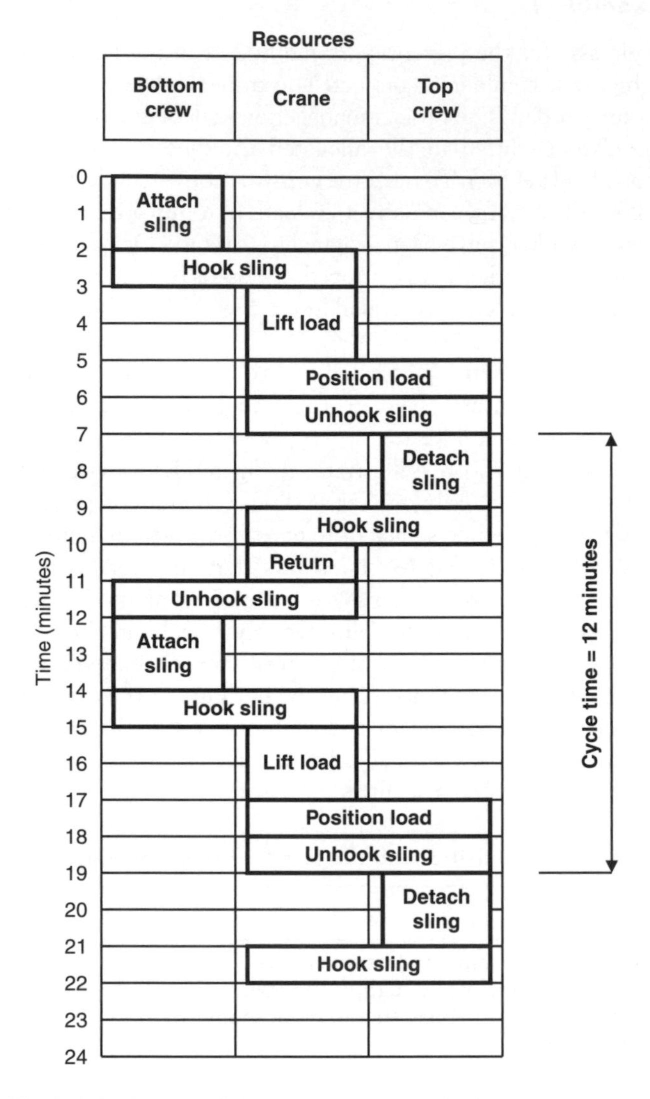

Figure 11.7 The MAC schedule of the current work method.

The cost of the present method is calculated as follows:

Crane	$5,000 × 3/8 hrs =	1,875.00
2 Crews of workers	2 × $1,000 × 3/8 hrs =	750.00
1 Set of slings	$100 × 3/8 hrs =	37.50
Total cost (3 hrs)		$2,662.50

The cost of handling one load of formwork is therefore $2,662.50/15 = $177.50.

An alternative work method must be capable of moving formwork faster and better utilising the committed crews of workers. The obvious solution is to increase the number of sling sets. Let's try to employ three sets of slings. Assume that two of the sling sets are located on the level where the bottom crew is situated while the third set is on the level where the top crew is placed at the start of the work. Alternatively, it could have been assumed that all three sets of slings are locked in the storeroom on the ground floor at the start of the day. The sequence of work would be the same for both assumptions, but the latter one would require a longer lead-time before reaching a regular cycle time of work.

The revised work method in the form of MAC is given in Figure 11.8.

The revised method of work is capable of lifting three loads of formwork in 24 minutes. It means that it meets (and exceeds) the required target of 20 loads. This method has also improved utilisation of the crews of workers from 40 to 50 per cent for the bottom crew and from 50 to 63 per cent for the top crew. The crane is now fully utilised.

The cost of the improved method is calculated as follows:

Crane	$5,000 × 3/8 hrs =	1,875.00
2 Labour crews	2 × $1,000 × 3/8 hrs =	750.00
3 Sets of slings	3 × $100 × 3/8 hrs =	112.50
Total cost (3 hrs.)		$2,737.50

The cost of handling one load of formwork is then $2,737.50/20 = $136.88.

Apart from delivering the required quantity of formwork material, the improved method offers better utilisation of the committed resources and a lower cost of handling.

11.5.2 Example 2

This example examines the current work method of delivering bricks and mortar by a single barrow hoist to typical floors of a high-rise building. The construction schedule requires at least 2,500 bricks to be laid each day. The ready-mix mortar is brought to the site in trucks and stored in suitable containers near the hoist on the ground floor. Bricks are also stacked on the ground floor near the hoist.

Two crews of workers have been assigned to this task. One crew comprising labourers is available on the ground floor with the task of loading wheelbarrows with bricks and mortar. Another crew is available on the working deck. It comprises bricklayers and labourers with the task of unloading wheelbarrows and building brick walls.

Bricks and mortar are transported using wheelbarrows. A wheelbarrow can be loaded with either bricks or mortar. The capacity of one wheelbarrow is 40 bricks.

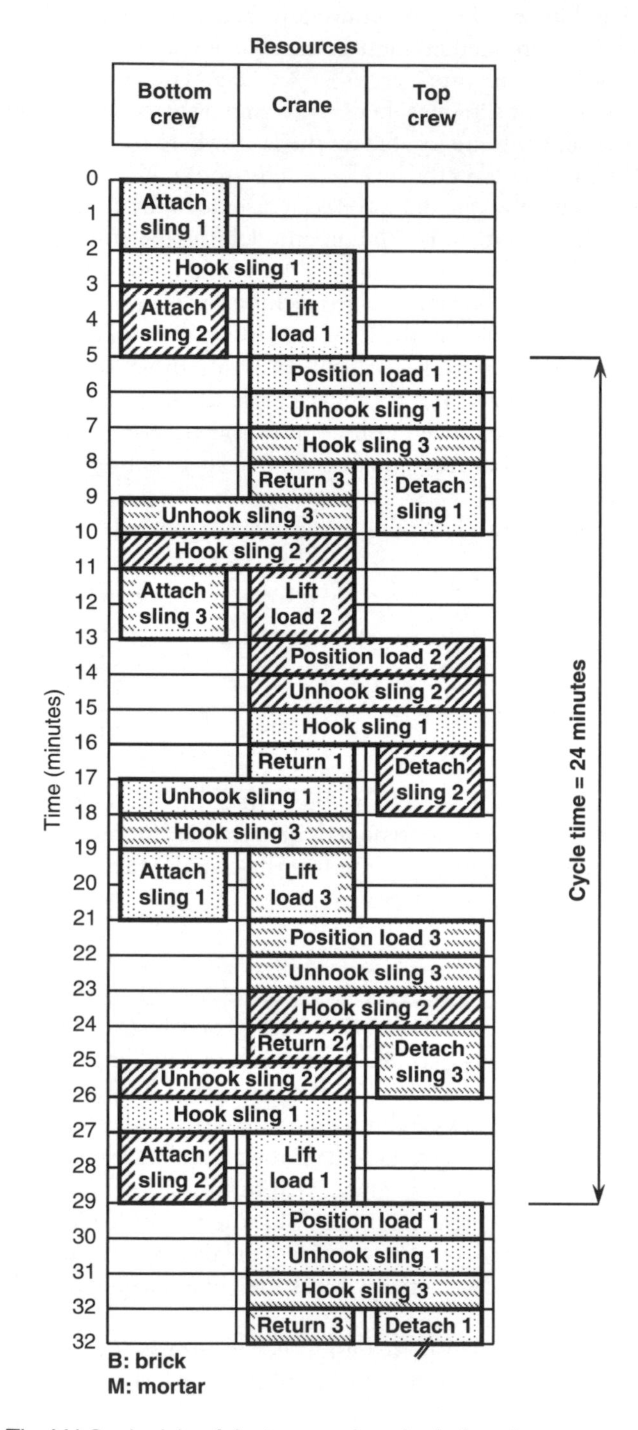

Figure 11.8 The MAC schedule of the improved method of work.

One cubic metre of mortar is required for every 80 bricks laid (or two wheelbarrows of bricks). It means that wheelbarrows of bricks and mortar are delivered in the ratio of 2:1.

A list of specific activities performed by the crews of workers including the standard times of work are given in Table 11.3. The cost of the committed resources is given in Table 11.4.

The present method of work is illustrated in MAC in Figure 11.9. It is based on the following decisions:

- One bricklayer lays approximately 500 bricks per day; consequently five bricklayers were employed to meet the daily production target of 2,500 bricks.
- One labourer is assigned to the ground floor to load wheelbarrows with bricks and mortar, and wheel them onto the hoist.
- One labourer is assigned to the working floor to take wheelbarrows off the hoist and unload them near the bricklayers.
- Only three wheelbarrows are available for moving bricks and mortar. They can be used interchangeably to move either bricks or mortar.

Table 11.3 Example 2 activities and their standard times of work

Activities	*Standard time (minutes)*
Fill barrow with mortar	1
Wheel and place mortar barrow on hoist	1
Take mortar barrow off hoist, wheel, empty and return to hoist	4
Fill barrow with bricks	2
Wheel and place brick barrow on hoist	1
Take brick barrow off hoist, wheel, empty and return to hoist	5
Hoist up	1
Hoist down	1
Change over barrows	1
Take empty barrow off hoist for refilling	1
Labourer to walk from stack of bricks to mortar storage	1

Table 11.4 Example 2 costs of committed resources

Resources	*Cost*
Bricklayer	$60/hour
Labourer	$40/hour
Hoist	$1,000/day
Wheelbarrow	$20/day

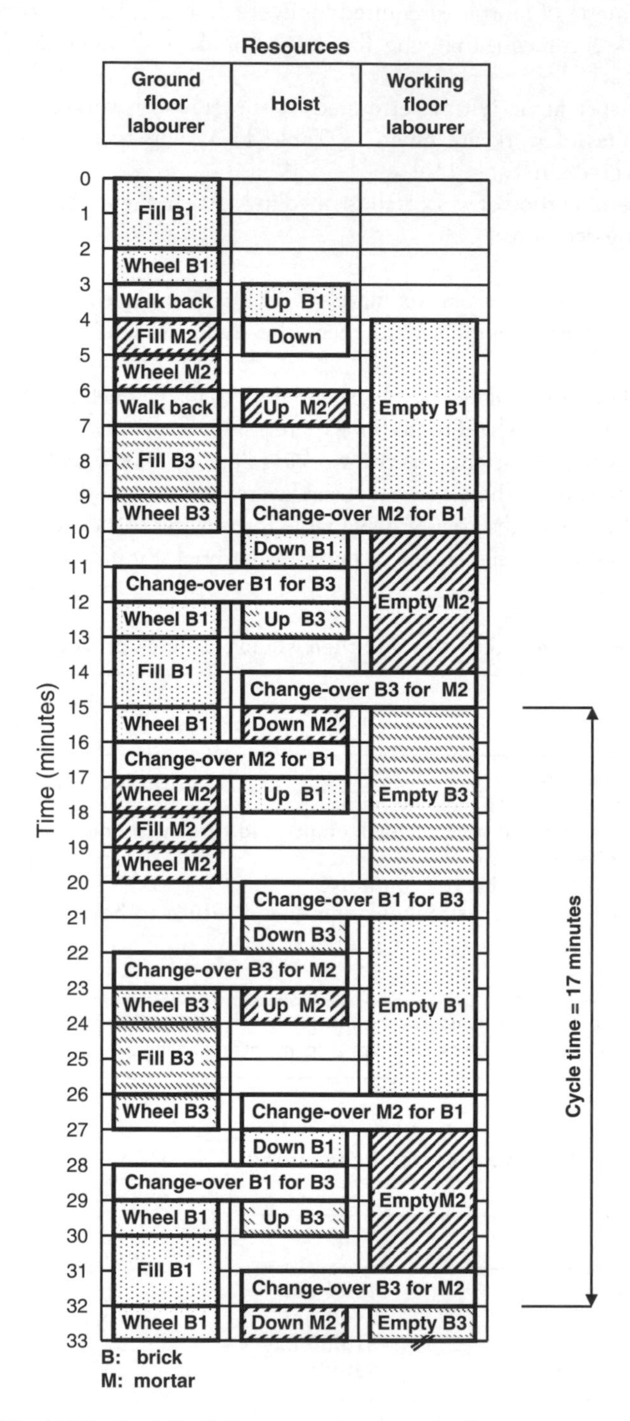

Figure 11.9 The MAC schedule of the present work method.

- Wheelbarrows are identified by numbers 1, 2 and 3. If a barrow carries bricks, its number is prefixed with 'B' and if it carries mortar, with 'M'.
- At the start of the working day all three wheelbarrows are located on the ground floor.

The present method of work delivers two wheelbarrows of bricks and one of mortar every 17 minutes. It means that 2,258 bricks are delivered to the bricklayers in eight hours. Since the required production output is 2,500 bricks per day, the present work method is inadequate. The cost of the current method of work is calculated as follows:

Hoist	$1,000 × 1 day =	1,000.00
5 Bricklayers	5 × $60/hr × 8 hrs =	2,400.00
3 Labourers	3 × $40/hr × 8 hrs =	960.00
3 Barrows	3 × $20 × 1 day =	60.00
Total cost (1 day)		$4,420.00

The cost of handling 1,000 bricks is (1,000/2,258 bricks) × $2,880 = $1,957.48

A closer examination of the MAC schedule in Figure 11.9 clearly shows that the labourer working on the working floor is already fully utilised, while the other resources have some spare capacity. By placing additional labourer on the working floor, the rate of progress is likely to increase. Let's see if it works. The revised MAC schedule based on two labourers placed on the working floor is given in Figure 11.10.

The improved method of work delivers two wheelbarrows of bricks and one of mortar every 14 minutes and thus 2,742 bricks per day, which is more than the scheduled production target of 2,500 bricks per day. The cost of the improved method is calculated as follows:

Hoist	$1,000 × 1 day =	1,000.00
5 Bricklayers	5 × $60/hr × 8 hrs =	2,400.00
3 Labourers	3 × $40/hr × 8 hrs =	960.00
3 Barrows	3 × $20 × 1 day =	60.00
Total cost (1 day)		$4,420.00

Although the method is capable of supplying 2,742 bricks per day, the employed bricklayers are able to lay only 2,500 bricks. Consequently, the cost of handling 1,000 bricks is (1,000/2,500 bricks) × $2,880 = $1,768.00.

The improved work method delivers sufficient quantity of bricks and mortar to meet the required daily output rate of 2,500 bricks. It also improves the cost of

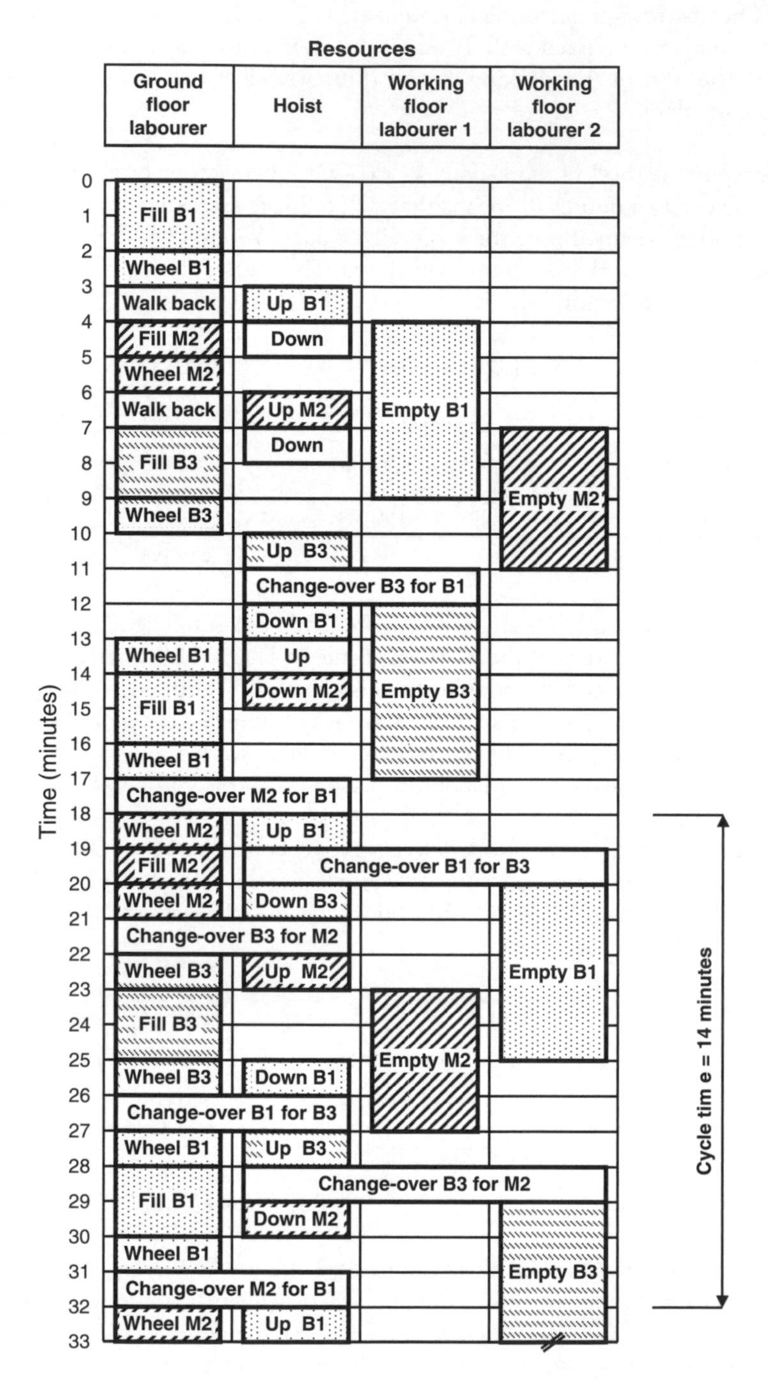

Figure 11.10 The MAC schedule of the improved work method.

handling bricks and mortar. But it does not effectively utilise two labourers employed on the working floor. The reader may consider developing other alternative solutions such as replacing wheelbarrows with a larger capacity plant.

11.6 Summary

This chapter examined the concept of work study, which is primarily a technique that aims at improving productivity of the production processes. It helps the manager to analyse work processes with a view to developing improved production methods. Through work measurement the manager is able to determine appropriate quantities of human and physical resources, measure their utilisation, and evaluate the economic viability of alternative methods of work.

Exercises

Solutions to the following exercises can be found on the following website: http://www.routledge.com/books/details/9780415601696/

Exercise 11.1

Calculate the level of utilisation of a window cleaner from the following information:

- Observed time for cleaning one window = 6 minutes
- Relaxation allowance + Contingency = 50% of Basic time
- Expected rating of the cleaner = 100
- Assessed rating of the cleaner = 120
- At the end of day 4, the cleaner cleaned the total of 160 windows. Assume that 1 working day is equal to 8 hours.

Exercise 11.2

Construction of a reinforced concrete high-rise building requires approximately 130 m^3 of concrete to be placed per floor. Concrete is delivered to the site in ready-mix trucks. The contractor has decided to discharge concrete from trucks to a kibble for lifting by a tower crane to the working floor. The contractor has one 1 m^3 capacity kibble on site but is able to obtain an additional kibble of 1.5 m^3 capacity on short notice, if needed.

The contractor has assigned one crew of labourers to load the kibble with concrete and the other crew to discharge concrete to the formwork. These crews of labourers are also responsible for hooking and unhooking the kibble to and from the crane as required. Table 11.5 shows a list of activities and their standard times expressed in minutes.

Table 11.5 Data for the work study example

Activities	*Standard time (min.) 1 m^3 kibble*	*Standard time (min.) 1.5 m^3 kibble*
Load kibble with concrete	0.5	1.0
Lift kibble	1.0	1.0
Manoeuvre kibble to work area	0.5	0.5
Unload concrete from kibble	1	1.5
Lower empty kibble to ground for refilling	1	1
Manoeuvre kibble to truck	0.5	0.5
Hook kibble to crane	0.5	0.5
Unhook kibble from crane	0.5	0.5
Lift crane hook (without kibble)	1	1
Lower crane hook (without kibble)	1	1

The delivery of materials to the site is restricted to between 8.30 am and 5 pm.

Using a MAC, illustrate the present method of work using a 1 m^3 capacity kibble. Will the contractor be able to place 130 m^3 in one working day? If not, try to improve the work method by using a 1.5 m^3 kibble.

Chapter 12

Risk and scheduling

12.1 Introduction

The common feature of scheduling techniques described in previous chapters was the reliance on the deterministic expression of time. Deterministic scheduling is based on the use of average time estimates in determining duration of activities. The planner who is responsible for the development of a schedule commonly adds allowances for uncertainty into average estimates of time. Top management then assesses the degree of 'uncertainty' that surrounds the entire project and formulates a project contingency. This contingency too is added to the schedule. Thus a contingency for uncertainty is often added to a schedule twice, first by the planner and second by top management. A rather risky estimate of duration of activities expressed as an average, as alluded to in Chapter 11, is thus transformed into a less risky estimate. In most cases contingencies are determined arbitrarily, largely based on past experiences of planners. Formulating activity durations in this manner does not permit the planner to understand the degree of exposure of a construction schedule to risk.

The deterministic approach to scheduling is clearly unscientific and incapable of realistically modelling the duration of construction projects. A better approach is to adopt the concept of probability scheduling.

Probability scheduling incorporates uncertainty into estimates of activity duration. This offers two specific benefits: it requires the planner and the planning team to study the project in detail in order to estimate the likely impact of identified risks on duration of individual activities; and it makes the use of contingencies largely redundant.

The purpose of this chapter is first to develop understanding of the concept of risk and how it can effectively be managed using the process of risk management, and second to relate it to probability scheduling.

12.2 Risk and uncertainty

Practically all non-trivial decisions and events in life have a wide range of possible outcomes, some expected, some hoped for and a few unforeseen. The possibility of unsatisfactory or undesirable outcomes occurring can never be ruled out. However, if the presence of uncertainty is predicted and its likely impact estimated with a reasonable degree of confidence, an appropriate management action may be taken to mitigate its likely impact. If, for example, the planner anticipates the likely delays in the supply of concrete to a project because of its location, the planner may mitigate the impact of this uncertain event by mixing concrete on site.

From the business point of view, the presence of risk is desirable since a natural balance exists between risk and opportunity. High-risk investments tend to pay proportionately larger premiums and, conversely, smaller returns are associated with low-risk investments. Risk and uncertainty are present in all aspects of construction work irrespective of size, complexity, location, resources or speed of building.

Uncertainty exists where there is an absence of information about future events, conditions or values. According to Porter (1981), uncertainty commonly gives rise to risk, which could be defined as an exposure to economic loss or gain, which has a known probability of occurrence. Uncertainty has an unknown probability of occurrence.

Most commercial decisions are made under conditions of uncertainty or risk. The presence of risk may not necessarily be a problem, particularly where its impact is low. And even if the impact is high, the planner may be able to develop a strategy for mitigating its impact or where possible use a greater level of risk to generate a higher level of income.

It is the goal of every planner to be able to perceive the presence of risk, and accurately predict its magnitude and likely impact on projects. This goal can only be satisfied through a systematic and disciplined approach to early identification, assessment and response to risk. Such an approach is commonly known as 'risk management'. Detailed information on risk management and its application to the construction industry can be found in Byrne and Cadman (1984), Cooper and Chapman (1987), Flanagan and Norman (1993), Edwards (1995) and Loosemore *et al.* (2006).

12.3 Principles of risk management

Risk management may be defined as 'the identification, measurement and economic control of risks that threaten the assets and earnings of a business or other enterprise' (Spence 1980: 22). It is a systematic way of looking at areas of risk and consciously determining how each should be treated. It is a management tool that aims at identifying sources of risk and uncertainty, determining their impact, and developing appropriate management responses.

The risk management process is defined by Australian/New Zealand Risk Management Standard AS/NZS 4360:2004 in terms of:

- Establishing the context
- Risk identification

- Risk analysis
- Risk response
- Monitoring and review.

Each component of the risk management process will now be examined in more detail.

12.3.1 Establishing the context

'Establishing the context' refers to examination of the organisation, its strengths and weakness, and the environment in which it operates. Other issues considered are identification of the stakeholders and interpretation of the organisation's business plan, including its goals and objectives. The process of establishing the context sets the scope and boundaries for an application of risk management and establishes the criteria against which risk is to be assessed. The purpose is to provide a logical framework for identifying, assessing and responding to risk.

12.3.2 Risk identification

The principal benefits of risk management are usually derived from the process of identifying risk. This is because identifying risk involves detailed examination of the project, its components and its strategy, which helps the project manager and the project team to understand better the complexity of the project, its design, the site on which it will be erected and the likely influence of a range of external environmental factors. The project manager thus becomes aware of potential weaknesses for which treatment responses must be developed. The project manager may also become aware of opportunities that may present themselves. But because the process of risk identification is usually a subjective one relying on the manager's ability to recognise potential risks, it may over- or under-emphasise, or even overlook, the importance of some risks.

Several methods of identifying risk are available. They fall into two distinct approaches: bottom-up and top-down.

Bottom-up approach to risk identification

Bottom-up risk identification works with the pieces and tries to link them together in a meaningful, logical manner. Examples of this approach are given below.

1 *A checklist approach*
   A checklist is a database of risks to which a firm has been exposed in the past. Depending on the volume of past data, it may be aggregated according to the type or size of project, the industry sector, the procurement type and the like. According to Mason (1973), this method offers the most usable risk

identification method for construction contracting by allowing the construction firm to identify risks to which it is exposed in a rational manner.

Reliance on historical data in a checklist may be both the strength and the weakness of this approach. As long as it is used only as a guide that helps the manager to identify risks in a thorough and systematic manner, a checklist approach may be highly effective. However, a fundamental limitation is that the project manager's look into the past may leave the project manager open to the bias of hindsight. The project manager may place too much emphasis on risks, which are either irrelevant as far as a new project is concerned, or have a real impact that is just too small to warrant any form of assessment and response.

2 *A financial statement method*
This method of risk identification is based on the premise that financial statement account entries serve as reminders of exposure to economic loss. Analysis of such statements would reveal the degree of exposure of resources. The weakness of the method is that it provides little help in identifying construction-related risk.

3 *A flow chart approach*
This approach attempts to construct a visual chart of the actual production process showing important components of the process and the flow of work. The project manager has an opportunity to focus on each element of the chart at a time and simultaneously consider the possibility of something going wrong with such an element. This approach may lead to identifying a series of risk events that may have significant impact on the project.

4 *A brainstorming approach*
Brainstorming is probably the most popular approach to risk identification. It involves project participants taking part in a structured workshop where they systematically examine every part of the project under the guidance of the workshop facilitator with a view to identifying likely risk events.

5 *A scenario-building approach*
This approach relies on the development of two scenarios: the most optimistic scenario where everything occurs as expected, and the most pessimistic scenario where everything goes wrong. The aim of this approach is to assess the two opposed scenarios in order to identify the factors or risks that might influence project performance.

6 *Influence diagram approach*
An influence diagram presents a more comprehensive view of the likely project risks. It helps to identify risk by a detailed assessment of cause–effect relationships among project variables. The project manager would first examine a particular variable outcome (effect) and by working backwards attempt to define the causes of variation. Once identified, these causes then become effects for which causes are sought in turn, and so on. For example, the cost escalation arising from variation orders may be caused by errors in design documentation, the client's changes to the scope, or new regulations

imposed by the local authority. The errors in design documentation may be caused by an incomplete brief, a lack of coordination or insufficient time set aside for design and documentation. The causes of the incomplete brief will further be examined, and so on. Graphically, the above simple example is illustrated in Figure 12.1, where circles or nodes represent high-risk events in the production process and arcs or lines illustrate cause–effect links between these events.

When all possible cause–effect relationships have been identified, the manager is presented with a highly detailed graphical map of possible risk events.

Top-down approach to risk identification

A top-down approach generates a holistic view of a project from which risk variables that are likely to have an impact on the project are deduced. There are two commonly used approaches:

1 *A case-based approach*
A case-based approach provides an opportunity to examine a specific case in its entirety. Most commonly, a past project for which a wealth of information is available would be selected as a case study. This examination of a case serves as an excellent training ground for new managers, who gain a holistic perspective of a typical project situation and its associated risks.

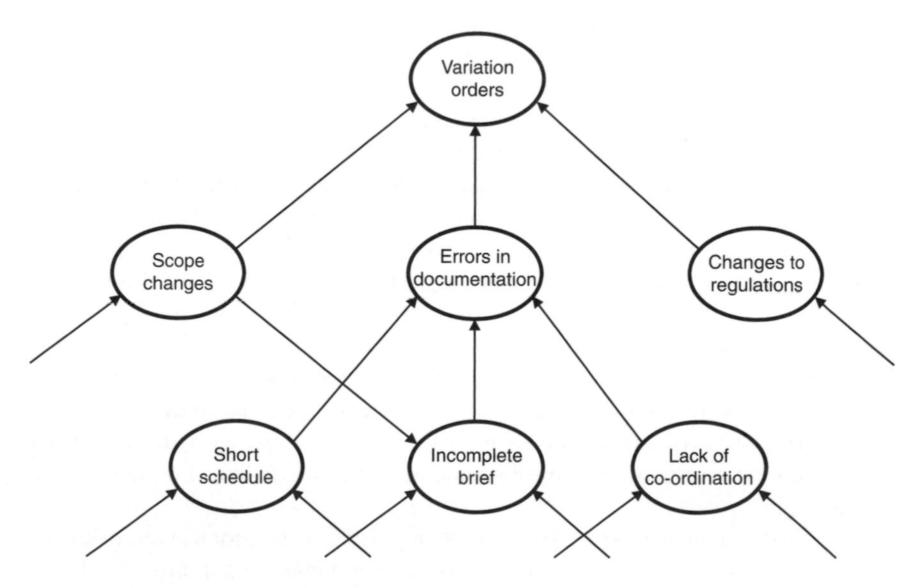

Figure 12.1 Example of an influence diagram.

2 *An aggregate or bottom-line approach*
A better-known term for an aggregate or bottom-line approach is a contingency allowance. It is a global perception of the volume of risk and its impact on the project. It is expressed as a percentage of a certain performance measure, such as contract period or project cost. Contingency is commonly formulated by top management and reflects subjective perception of the likely impact of risk on the project. While attractive for its simplicity, it lacks the ability to explain in any meaningful way the basis for the developing a risk management response to anticipated problems.

Before identified risks can be analysed, their likely magnitudes need first to be determined. This process is referred to as 'data elicitation'. Eliciting data for the identified risks is perhaps the most difficult task in risk management. Data are usually derived from databases, random experiments or the knowledge of experts.

Information extracted from a database or a random experiment provides 'objective data' that can be expressed in the form of a probability distribution. In this form, data are highly suitable for the quantitative risk assessment. Betts (1991), Ivkovic (1991) and Townley (1991) believe that objective data are preferred because of their consistency and perceived accuracy. But the use of databases in the construction industry is rare and the cost of random experiment high.

'Subjective data' are elicited by brainstorming, which accesses the knowledge and experience of the project participants. It may appear to be an arbitrary procedure or guesswork, but it is not so. If carried out by a properly structured brainstorming process, managed by an experienced facilitator, it provides collective expert knowledge of high quality. The best-known brainstorming technique is the Delphi method, which seeks information from a group of experts by means of an iterative questionnaire technique (though, while highly effective, the Delphi method can be rather time-consuming).

Subjective data may be expressed as likelihood and consequence of risks in 'high' or 'low' terms (see *Qualitative risk analysis* in section 12.3.3), or as estimates of specific values of risk variables that fit simple probability distributions, such as uniform, triangular, trapezoidal and discrete (see Figure 12.2). The simplest way of transforming subjective data into probability distributions is to arbitrarily determine a two-point estimate of the highest and lowest value of the random variable. These two points describe the uniform probability distribution, which is the simplest probability distribution. A three-point estimate of the highest, most likely and the lowest values describes the triangular probability distribution. An extension of the triangular distribution is the trapezoidal distribution, which is characterised by two estimates of the most likely values together with the highest and the lowest estimates. Shapes of simple probability distributions are given in Figure 12.2.

Uniform distribution assumes that a range of possible values for a given risk variable can only be expressed between its minimum and maximum limits. While the use of this distribution may be acceptable in some applications, for example in

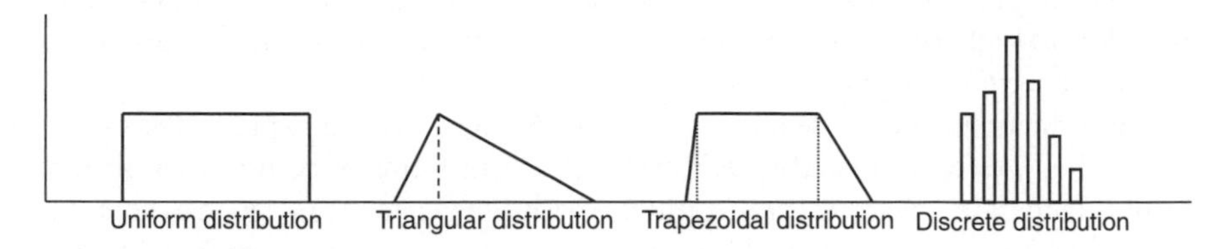

Figure 12.2 Examples of simple probability distributions.

predicting the exchange rate of the currency, it is doubtful that it would accurately model construction cost and time.

Triangular distribution is viewed as being adequate for most applications, particularly in estimating cost and time (Wilson 1984; Raftery 1990). It is characterised, in addition to its minimum and maximum limits of values, by the most likely value.

Trapezoidal distribution is described by the minimum and maximum limits, and by two estimates of the most likely values.

Discrete distribution shows frequency of occurrence of various outcomes of a given risk variable. Such frequencies are in fact probabilities of occurrences that describe the range of possible outcomes.

When using simple probability distributions, care is required in determining the lower and upper limits of the distributions. There are no precise rules governing the determination of these limits; however, they should be set on the assumption that there is only a small chance, for example 1–2 per cent, that their values would be exceeded (Hertz and Thomas 1983; Wilson 1984).

12.3.3 Risk analysis

The purpose of risk analysis is to measure the impact of the identified risks on a project. Depending on the available data, risk analysis can be performed qualitatively or quantitatively.

Qualitative risk analysis

Qualitative assessment of risk is popular for its simpler and more participative approach. It involves subjective assessment of the derived data in terms of risk likelihood and consequence. When appropriately structured and systematically applied, qualitative risk analysis serves as a powerful decision-making tool.

'Internal Audit and Risk Management Policy for the NSW Public Sector' of the NSW Government (2009) and the Australian and New Zealand 'Risk Management Standard 4360' (AS/NZS 2004) describe a qualitative risk assessment process in which risk events are expressed in terms of likelihood and consequence. The aim of this form of risk assessment is to isolate the major risks and exclude those that are regarded as minor or have low impact.

An estimate of the likelihood of each risk may be expressed on a simple scale from low to high, or on a more detailed scale from rare, unlikely, moderate, likely, to almost certain (AS/NZS 2004).

An estimate of the consequences of each risk may also be expressed as either low or high, or as suggested in AS/NZS (2004) in terms of being insignificant, minor, moderate, major or catastrophic. A risk management matrix can then be formed using estimates of likelihood and consequence. An example of such a matrix is given in Figure 12.3.

In Figure 12.3 minor risks will commonly be ignored. Moderate risk will be carefully analysed and appropriate management responses formulated, while major risk will be given the utmost attention.

Quantitative risk analysis

Quantitative risk analysis techniques are stochastic or random in nature. The most important feature of a stochastic process is that the outcome cannot be predicted with certainty. The application of any such technique requires that risk data be expressed as a probability distribution.

The choice of technique usually depends on the type of problem, the available experience and expertise and the capability of the computer software and hardware. A wide range of techniques is available including sensitivity analysis, probability analysis using Monte Carlo simulation, decision trees, utility functions, and full control interval and memory (CIM) analysis (Cooper and Chapman 1987). A brief review of sensitivity and probability analyses is presented.

1 *Sensitivity analysis*

Sensitivity analysis seeks to place a value on the effect of change of a single risk variable within a particular risk assessment model by analysing that

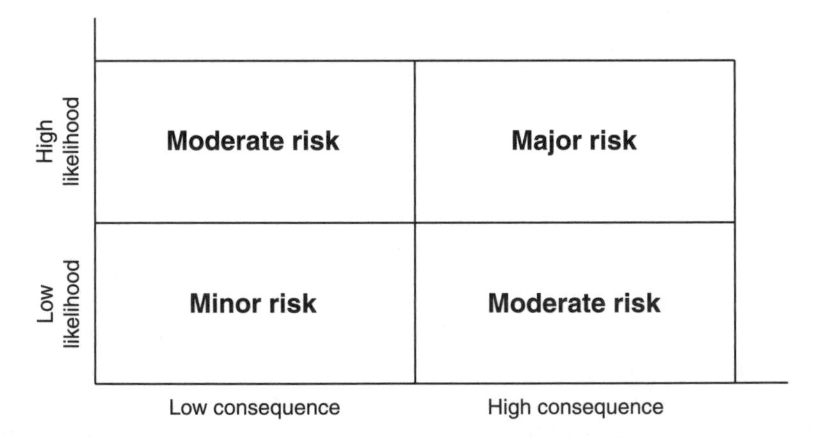

Figure 12.3 A simple risk management matrix (adapted from NSW Government (2009)).

effect on the model output. A likely range of variations is defined for selected components of the risk assessment model and the effect of change of each of these risk variables on the model outcome is then assessed in turn across the assumed ranges of values. Each risk is considered individually and independently, with no attempt to quantify probability of occurrence.

The importance of sensitivity analysis is that often the effect of a single change in one variable can produce a marked difference in the model outcome. Sometimes the size of this effect may be very significant indeed.

In practice, a sensitivity analysis will be performed for a large number of risks and uncertainties in order to identify those that have a high impact on the project outcome and to which the project will be most sensitive. If the manager is interested in reducing uncertainty or risk exposure, then sensitivity analysis will identify those areas on which the effort should be concentrated.

The outcome of sensitivity analysis is a list of selected variables ranked in terms of their impact. This is best presented graphically, with a spider diagram being the most preferred form of representation. A spider diagram is an X–Y graph with the horizontal and vertical axes showing a percentage change in the outcome variable and the percentage of change in the level of risk respectively. A spider diagram in Figure 12.4 illustrates assessment of the impact of four risk variables:

i Problems associated with rock excavation
ii Design changes

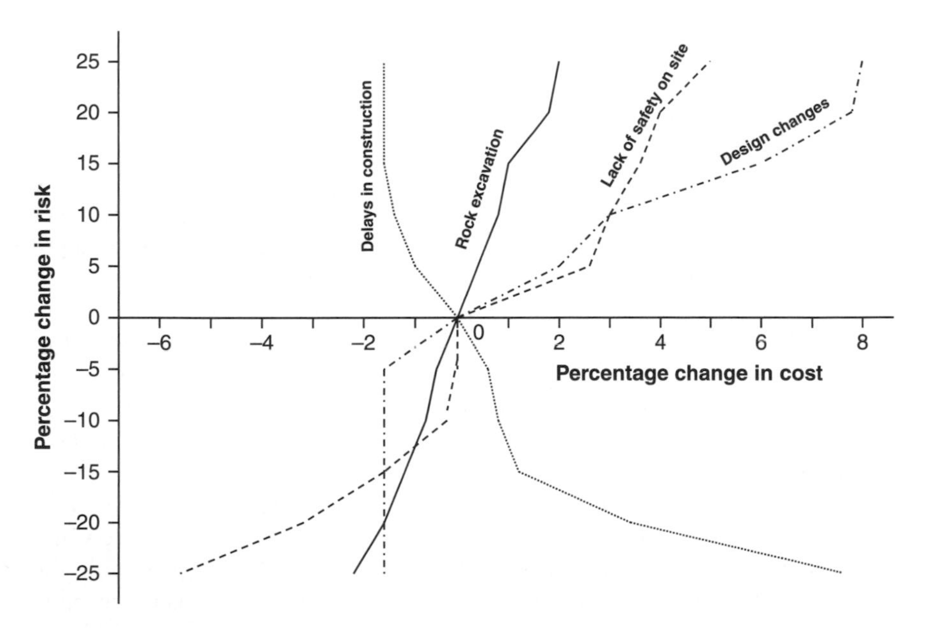

Figure 12.4 A spider diagram of the four risk variables.

iii Lack of safety on site
iv Delays in construction.

For each risk variable, the diagram graphs the impact on the cost of a defined proportionate variation in a single risk variable that has been identified as having some risk associated with its cost estimate.

Assume that the risk in the first three defined variables is related to cost overruns while the risk in the fourth variable affects the rate of production. Assume further that the risk could increase or decrease by 5, 10, 15 or 20 per cent. Each risk variable is analysed separately for different increments of risk. The resulting effects on the total project cost is determined and plotted as a series of lines.

It is clear that the flatter the line, the more sensitive the cost variation in that variable. 'Design changes' is the most sensitive variable closely followed by 'Lack of safety on site'. The impact of 'Delays in construction' becomes severe when the percentage variation in the variable is more than 20 per cent.

The major weakness of sensitivity analysis is that selected risk variables are treated individually as independent variables, and interdependence is not considered. The outcomes of sensitivity analysis thus need to be treated with caution where the effects of combinations of variables are being assessed.

2 *Probability analysis*

The weakness of sensitivity analysis is that it looks at risks in isolation. Probability analysis overcomes this problem. Probability analysis is a statistical method that assesses a multitude of risks that may vary simultaneously. In combination with Monte Carlo simulation, it provides a powerful means of assessing project uncertainty.

The key to probability analysis is the development of a risk analysis model. The model should include variables affecting the outcome, taking into account the interrelationships and interdependence between the variables. Each risk variable in the model is expressed as a probability distribution. The model must permit assessment of risks at the desired level of detail and accuracy. This may require breaking down risk variables into their subcomponents in order to accurately assess the impact of risk on such subcomponents.

Probability analysis is commonly performed using Monte Carlo simulation. The Monte Carlo technique generates random numbers that are related to probability distributions of individual risk variables in the model. After each iteration, the model calculates one specific outcome from the generated risk values from individual probability distributions. After a large number (at least 100) of iterations have been performed, the Central Limit Theorem, an important concept in statistics, ensures that the outcomes fall on a normal curve, which is fully described by its mean and standard deviation. Monte Carlo simulation will be examined in more detail later in this chapter while the Central Limit Theorem will be discussed in Chapter 13. A typical probability analysis process using Monte Carlo simulation is illustrated in Figure 12.5.

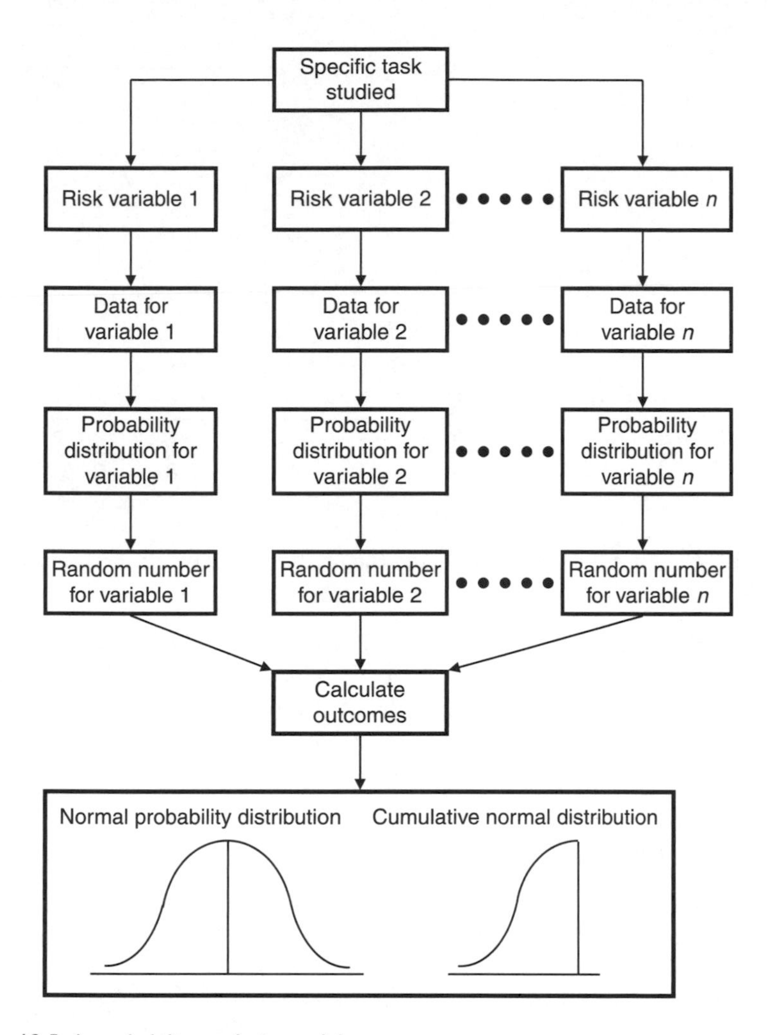

Figure 12.5 A probability analysis model.

The outcome of probability analysis is a normal distribution expressed by its mean and standard deviation. For example, assume that the result of probability analysis is the mean of the cost estimate of \$22,000 and the standard deviation of the cost estimate of \$1,120.

Given the properties of the normal distribution, there is a 68 per cent probability (the area between +1 SD to −1 SD from the mean) that the project cost lies in the range between \$20,880 and \$23,120, and a 95 per cent probability that it lies in the range between \$19,760 and \$24,240 (refer to Figure 12.6(a)). In cumulative terms, approximately 16 per cent of the area under the normal curve lies to the left of −1 SD and 84 per cent to the left

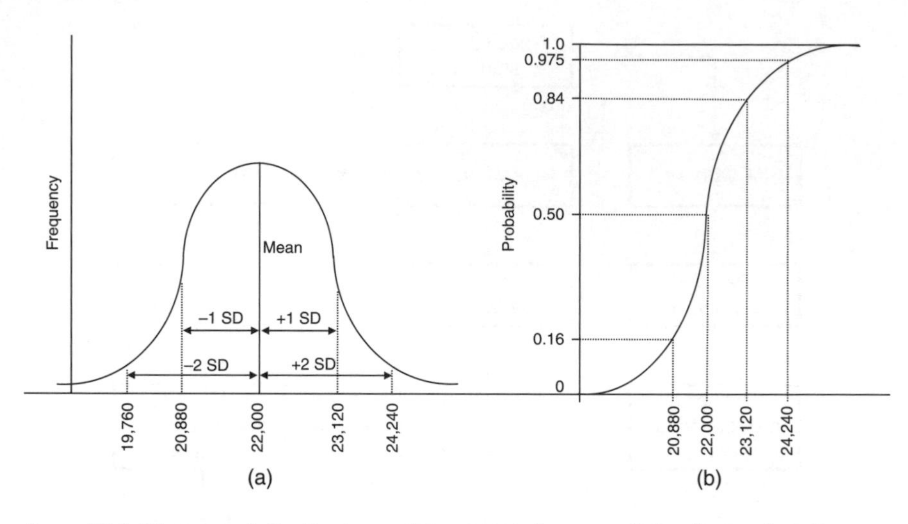

Figure 12.6 The normal distribution and the cumulative normal distribution functions.

of +1 SD. The similar values for −2 SD to +2 SD are 2.5 per cent and 97.5 per cent respectively. The likely cost outcomes for various probabilities can easily be read off of a cumulative normal distribution curve given in Figure 12.6(b). For example, there is 84 per cent probability that the cost will be less than \$23,120 and 97.5 per cent probability that it will be less than \$24,240.

12.3.4 Risk response

Risk response is an action or a series of actions designed to deal with the presence of risk. This involves developing mitigation and treatment strategies and implementing them.

The manager may adopt one of two possible response strategies: transfer the risk and/or control it.

Risk transfer

Risk transfer involves shifting the risk burden from one party to another. This may be accomplished either contractually by allocating risk through contract conditions or by insurance. Contractual transfer is a popular form of risk transfer. It is used extensively in construction contracts, where one party with power transfers the responsibility for specific risks to another party. The most commonly occurring risk transfers involve:

- Clients transferring risk to contractors and designers
- Contractors transferring risk to subcontractors.

Risk transfer by insurance is highly desirable in those situations where insurance policies exist. The purpose of insurance is to convert the risk into a fixed cost. This approach assigns a dollar value to the risk.

Risk control

When risk can neither be transferred nor insured, management action is required to manage it. This is commonly achieved through processes of risk avoidance and risk retention.

A simple approach to managing risk is to avoid it in the first place. For example, a risk of unreliable concrete supply from a single supplier can be avoided by engaging two suppliers. Similarly, if the contractor believes that the client imposes an excessive burden of risk through contract conditions, the contractors may avoid the risk altogether by not bidding for the job.

Not all risks can be avoided. However, their impact may potentially be reduced, for example by developing alternative solutions or even redesigning sections of the project. In collaboration with the project stakeholders, the manager needs to develop appropriate treatment strategies and to assign responsibility for their implementation.

While risk avoidance and risk minimisation can help reduce the overall level of risk, some residual risk will always remain. It is this risk that requires close attention. The most common approach to controlling residual risk is to convert it into a contingency allowance. Contingency may be expressed as a single-value estimate of a particular measure of performance such as time, cost, hours of work, etc., or as a probability distribution of a particular measure of performance.

Perhaps the most serious weakness of a single-value contingency approach is its inflationary impact on the base estimate. When a percentage contingency is added to the base estimate, every item in that estimate will be inflated by the percentage figure of the contingency, irrespective of whether such items represent 'risk' or not. For example, a 10 per cent contingency added to the cost estimate inflates the cost of each item in the estimate by 10 per cent. If the Pareto Principle holds, then only about 20 per cent of such cost items in the cost estimate represent the real risk. They should then account for approximately 80 per cent of the uncertainty.

When a risk contingency is expressed in the form of a probability distribution, it is possible to determine its value at a specific level of probability. For example, a prudent manager may prefer to accept a certain level of risk but at the probability level of 85 per cent or higher. It is then a simple task to determine the actual value of the risk at that level of probability.

12.3.5 Monitoring and review

Although listed last, monitoring and review is not the end step of the risk management process. In fact monitoring and review starts almost from the beginning and is maintained throughout the entire risk management process. Since no risk is static, monitoring and review ensures that changing circumstances are identified and reassessed. This is necessary to ensure relevance of the entire risk management process.

12.4 Risk management plan

A risk management plan is a written statement describing the entire risk management process. It specifies the processes employed and summarises the results obtained in each stage of risk assessment. It aggregates identified risks into low–high impact and likelihood types. It lists elicited values of risk variables and assigned probability distributions, where appropriate. It defines treatment strategies particularly for high impact or likelihood risks and assigns responsibilities. It lists discarded risks. It also reports on resource requirements, timing of actions, and monitoring and reporting mechanisms.

12.5 Risk-time contingency calculation techniques

Uncertainty refers to an event which is indefinite, indeterminate, not certain to occur, problematic, not known beyond doubt and/or not constant. The risk involved with construction projects can never be eliminated, although the amount of uncertainty can be estimated and allowed for when calculating construction schedules and then risk can be effectively managed. This section looks at some of the ways of quantifying and allowing for uncertainty in the form of a time contingency in CPM schedules. In relation to time contingencies Horowitz (1967) believes that time contingencies should not be added to activity duration estimates when initially calculating activity durations, but rather should be applied to the activities that are likely to be susceptible to a delay. For example, a time contingency for bad weather should only be added to activities that are likely to be affected by bad weather. The amount of the time contingency for bad weather should be a proportion of the total amount of expected bad weather days over the entire project duration. Healy (1999) assumes that the optimum safety allowance on critical path based activities is 30 per cent of the original duration estimate. Whilst there is no definitive way of allowing for a time contingency in CPM schedules, the most common ways of allowing for time contingencies include:

- Add time contingencies to risky activities
- Add a time contingency to the whole project as a lump-sum allowance, usually inserted succeeding the final activity and prior to the terminal activity
- Break up the lump-sum time contingency into a number of smaller contingencies that are then added to the schedule at regular intervals
- Randomly allocate 'non-work days' in the project schedule calendar.

12.5.1 Calculating time contingencies for critical path schedules

The risks that a contractor would usually allow for in the time schedule are risks that fall within the scope of works as defined in the particular project contract. As there are many forms of contract, each with its own level of risk allocation between contracting parties, it is common practice for the contractor to seek legal advice to interpret the particular contract type and identify the risks to be borne by the contractor prior to commencing the project. Some of the risks that the contractor would allow for and add to a time schedule in the form of a time contingency include:

- Inclement weather
- Contaminated or hazardous materials
- Industrial disputes
- Design changes
- Other.

Project planners usually allow for a time contingency in the form of a percentage of the overall project duration. An example of the breakdown of a time contingency is shown below:

- Inclement weather (7 per cent)
- Contaminated or hazardous materials (1 per cent)
- Industrial disputes (2 per cent)
- Design changes (1 per cent)
- Other (2 per cent)
- *Total time contingency – 13 per cent (of the net project duration).*

The percentage allowance for each of the above components will vary from project to project and is dependent on the project particulars. A commonly used method for calculating a time contingency for 'Inclement weather' is described in the following steps:

1 Identify the inclement weather conditions which are excluded from 'Extension of time claims' within the project contract, for example:

   i. Rain
   ii. Heat
   iii. Wind.

2 Identify the project geographic location and work climate, for example:

   i. Sydney, New South Wales, Australia
   ii. New six-storey commercial building open to all weather elements.

3 Review local weather station data and ascertain the average percentage of delay days experienced due to inclement weather over the last, say, 10 years:

 i Rain:

  a Calculate the mean number of days where rain exceeded 2.5 mm per month, say 4 days
  b Determine the number of days where rain may affect work (i.e. until the building is 'watertight'), 6 (month duration) at 4 day/month = 24 days
  c Calculate the number of days in the schedule, say 330 calendar days
  d Determine the average percentage of delay days on the schedule, 24/330 = 7%.

 ii Heated-off days: by applying the above steps, determine the average number of days where the temperature exceeded 36°C, say (1 per cent)
 iii Wind: again, by applying the above steps, determine the average number of days when wind caused work stoppages, say (1 per cent).

4 Graph the results. See Figure 12.7.

12.5.2 Monitoring and re-forecasting time contingency allowance

It is not enough to simply allow for a time contingency in a project schedule. The time contingency needs to be monitored, re-forecast and reported as part of the regular project reporting process. The correct way to use the delay contingency is to record the delays incurred, re-forecast or update the construction schedule taking

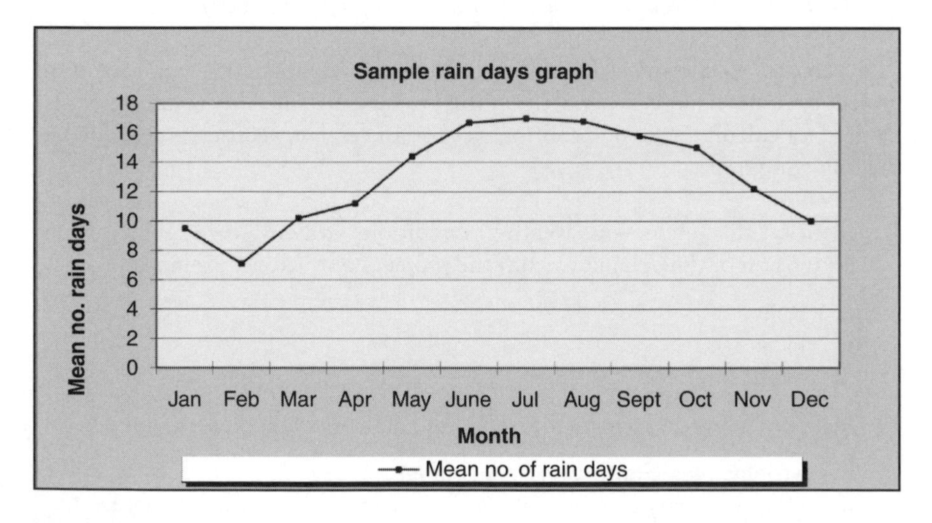

Figure 12.7 Rain days graph.

into account the delays incurred and report on the rate of delay days being expended as well as the delay days remaining over the balance of the project. This will enable the planner to report on any likely shortfall of delay days later in the project.

The steps to ensure that the time contingency is monitored, re-forecasted and reported include:

1 Record delays in a delay register stating time of delay, duration of delay and reason for delay
2 Status the construction schedule by deducting delay days incurred from the overall time contingency
3 Compare the delays incurred to date against the delay days remaining, based on this result forecast the final delay days forecast to remain at project completion (see Figure 12.8).

The delay tracking process can also highlight the project productivity. This can be done by looking at the work days completed to date (actual days worked) *divided by* the project days that have passed *less* the delay days expended to date.

12.6 Probability scheduling

The traditional deterministic process of scheduling, which includes the formulation of time contingencies, has already been discussed in this book. Attention was drawn to the practice of adding two layers of contingencies to average estimates of activity durations in a highly subjective manner. Clearly, the deterministic approach with arbitrarily added contingencies is deficient.

An alternative approach is to apply the process of risk management to the entire scheduling process. This requires identification of risk that is likely to impact on individual activities in a schedule, its assessment and the development of treatment strategies for minimising its intensity. With this information in hand, the planner together with other project 'experts' ascertains the extent of variability of activity duration and expresses it as a probability distribution. The combined impact of risk on the schedule is then assessed using probability analysis. The planner is able to predict, with a high degree of confidence, that a project will be completed by a certain date. This approach is much more robust, and in comparison with single-value or deterministic scheduling, it should produce more accurate schedules.

Some people argue that since probability scheduling depends largely on subjective assessment of risk variables and subjective expression of their values, this approach is inaccurate and in no way superior to single-value scheduling. Keeping in mind the probability scheduling approach and its features described above, let's compare it with the traditional, single-value scheduling. Single-value scheduling is mainly a mechanical process of establishing durations of activities within the logic of a construction schedule. For a preferred construction strategy, the planner assigns activity durations based on the planner's personal experience or industry standards in the form of mean values. Mean values of activity durations may or

Planning Department

PROGRESS REPORT

Project Sample Project

Progress Report On: Area 1

Programme Ref: E022

for progress monitored up to: 12/12/2010

Progress reported by: AZ

Programme Data:

Nett Time Allowance: 110 Days
Delay Allowance: 44 Days = 40 % of Nett Time
Total Time: 154 Days

Estimate at Completion:

Nett Days: 30
Delay Days: 10
Total Days: 131

Start Date: 16/01/2011 **Completion Date:** 5/10/2011 **Projected Completion Date:** 5/10/2011

Delay Category	Programme Allowance		Delays Incurred (to the critical path)		Delay Allowance to Date		Extensions of Time to Completion Date			Estimate at Completion			Remarks
	Days	% fo Nett	Days	% fo Nett Time Exp.	Days	% fo Nett Time Exp.	Days Claimed	Days Approved	Likely Approvel	Days	% of Allow	% of Nett	
Inclement Weather	44	40%	8	9.00%	28	31.10%	1	0	0	7.5	17.00%	6.80%	
Industrial Action													
Design													
Documentation													
Variations													
Client, Authorities, etc													
Other:										2.5	5.70%	2.30%	
Other:													
TOTALS	44	100	8	9.00%	8	31.10%					22.70%	9.10%	

		(a)	(b)	
Total Time Expended		97 day/s (c)	= ____	% of total time
Nett Time Expended	(c – a)	89 day/s (d)	= ____	% of total nett time
Nett Time Achieved		64 day/s (e)	= ____	% of total nett time
Nett Status	(e – d)	-25 day/s (f)		
Delay Allowance Status	(b – a)	20 day/s (g)		
Overall Status	(f + g)	-5 day/s (h)		

Critical Path Assessed On:

Figure 12.8 Monitoring of time contingency allowance.

may not be representative of actual durations. If a schedule is developed and the level of contingency determined without first applying a rigorous process of risk management, it must be asked whether the planner and the firm's top management really have detailed knowledge of the project, understand its main features and complexities, and are aware of the extent of exposure to risk. It follows that probability scheduling is a more robust approach that should make the planner feel more confident of formulating the best possible scheduling strategy.

Probability scheduling is commonly performed either by Monte Carlo simulation or by a PERT technique. The former approach will now be briefly discussed. PERT will be examined in detail in Chapter 13.

12.6.1 Monte Carlo simulation

Simulation attempts to predict in advance the outcome of a certain decision. It implies 'having a look before a leap'. Monte Carlo simulation is a method of using random numbers to sample from a probability distribution where random numbers are related to the relative probabilities (frequencies) of the factor being simulated so that the more probable values of the factor are picked appropriately more often. Through sampling and with a sufficient number of iterations, the Monte Carlo method will recreate the input distributions. It is recommended that at least 100 iterations be performed in order to minimise a sampling error.

The following simple example demonstrates the working of Monte Carlo simulation. Let's assume that the frequency of sales of computers achieved by the computer manufacturing company over the past 100 days is given in the first two columns of Table 12.1. The third column contains frequencies of sales expressed as probabilities.

The frequency of sales of computers in the past 100 days is expressed as a discrete probability distribution. To ensure that simulation accurately represents the

Table 12.1 The frequency of sales of computers

Computers sold per day	*Frequency of sales*	*Probability (%)*	*Random numbers*
20	2	1	0
21	7	4	1–4
22	26	16	5–20
23	40	24	21–44
24	22	13	45–57
25	19	12	58–69
26	26	16	70–85
27	10	6	86–91
28	8	5	92–96
29	4	2	97–98
30	2	1	99
Total	166	100	

frequencies of sales, consecutive random numbers that are equal to the value of the probabilities will be assigned to each frequency group. Since total probability is equal to 1 or 100 per cent, in total, 100 random numbers from 0 to 99 are used.

The probability of selling 20 computers is 1 per cent, which is equivalent to one random number. Let's assign the first random number in the series '0' to represent this probability.

The probability of selling 21 computers is 4 per cent. Therefore four random numbers 1–4 are assigned for this probability. The next probability value of 16 per cent, which reflects the sale of 22 computers. It is assigned the next 16 consecutive numbers, 5–20, and so on until all 100 random numbers have been assigned. Since any random number is equally likely to occur, the more probable frequency groups will occur more often in the simulation process provided it has been iterated a large number of times.

A discrete probability distribution of the sales of computers can easily be constructed and simulated in a spreadsheet enhanced with an add-in risk analysis software such as @Risk. After 100 or so iterations, the simulated frequencies of sales begin to stabilise to closely match those of the input distribution.

12.6.2 Monte Carlo scheduling

The technique of Monte Carlo simulation is commonly used in probability scheduling. The heart of probability scheduling is a computer-generated critical path schedule, which either has simulation capabilities or could be enhanced with such simulation capabilities through add-in simulation software. The 'iThink' software is an example of the former case and the 'Monte Carlo for Primavera' of the latter.

The development of a probability CPM schedule requires the activity's duration to be expressed as a distribution of possible durations. Triangular distribution is the one most commonly adopted. It expresses values of durations as:

- Most optimistic
- Most pessimistic
- Most likely.

After all activities in a schedule have been expressed as probability distributions, activity-to-activity, resource-to-resource and activity-to-resource correlations will be set by the planner to create a highly realistic scheduling model.

A risk analysis is then performed on a schedule. In individual iterations, the input duration distribution of each activity is sampled and ESD, EFD, LSD and LFD (see Chapter 3), total duration and float values are calculated. These values will vary from iteration to iteration. When the value of total float is zero or even negative, the activity is critical. Because durations of activities vary for each iteration, the position of a critical path is not constant. The computer software will calculate in percentage terms the criticality of each activity. For example, it may show that the activity 'formwork to walls' is on the critical path 67 out of 100 iterations.

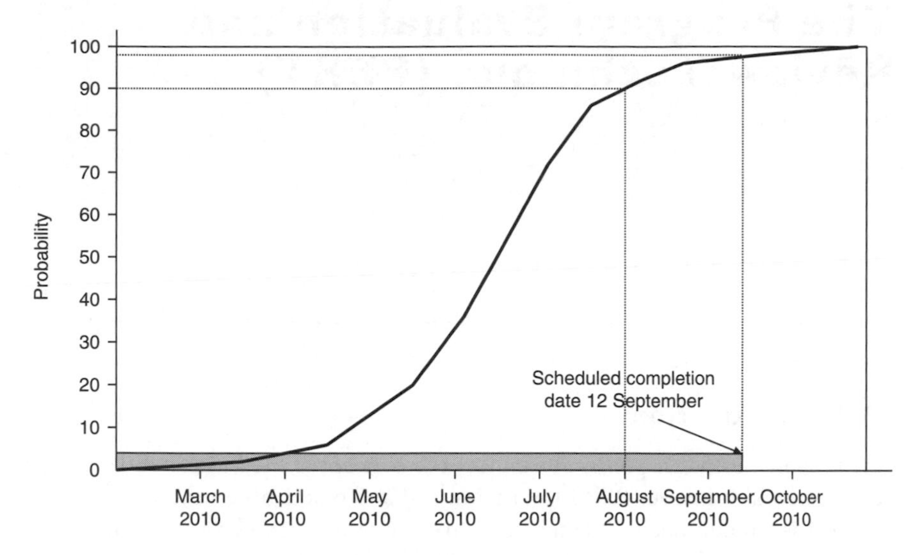

Figure 12.9 The cumulative density function of the project duration.

The output distribution of a simulated schedule is, according to the Central Limit Theorem, a normal distribution. When expressed as a cumulative density function, the planner is able to determine the probability of completing the project by the contract date. This is illustrated in Figure 12.9. The probability of completing the project by the contract date of 12 September 2010 is about 98 per cent. Similarly, there is about 90 per cent probability that the project could be completed by 1 August 2010.

Some simulation software is also able to express activity costs as probability distributions from which the software generates a cumulative cost-density function. The planner is then able to determine the probability of completing the project within the contract cost. By comparing both the cumulative time and the cumulative cost-density functions, the planner gains better understanding of the likely project performance and is able to formulate a much more representative level of contingency.

12.7 Summary

This chapter introduced the concept of risk and uncertainty into scheduling, and methods of allowing for risk as a time contingency in construction schedules. First it defined risk and uncertainty, then the process of risk management. In the latter part of the chapter, probability scheduling using Monte Carlo simulation was briefly described. The second method of probability scheduling known as PERT will be examined in the next chapter.

Chapter 13

The Program Evaluation and Review Technique (PERT)

13.1 Introduction

The purpose of this chapter is to examine the concept of the Program Evaluation and Review Technique (PERT) in theoretical and practical terms.

PERT is a technique that can be used to plan and control projects surrounded by uncertainty. The fundamental aim of PERT is to track the progress of a project and show, at different time intervals, the probability of completing that project on time. Unlike Monte Carlo-based critical path method (CPM) scheduling, which relies on sampling, PERT relies on statistics, particularly the Central Limit Theorem, in calculating the probability of project outcomes.

PERT was developed in 1958, in parallel to the CPM, by Malcomb *et al.* (1959) to assist the US Navy in the development of the Polaris submarine/ballistic missile system. It was primarily conceived as a project control technique to ensure that the highly complex and strategically important Polaris project would be delivered on time. The fact that the project was completed and commissioned well ahead of its schedule is being largely attributed to PERT.

Throughout the 1960s PERT remained the principal planning and control technique in the United States, particularly on large military and aerospace projects such as Atlas, Titan I, Titan II, the Minuteman rocket systems and the Apollo space programme. However, by the late 1960s the CPM gained in popularity and has since become the most widely used planning and control technique.

The lack of off the shelf software support is probably the main reason why PERT is rarely used in the construction industry. Nevertheless, the concept of PERT is fundamentally sound and suitable for planning and control of construction projects. It provides a plausible alternative to a Monte Carlo-based scheduling approach.

This chapter will first examine the original event-oriented modelling in PERT, which is based on the arrow network. For this approach, the probability concept in PERT will be defined and the computation method developed. Later, the PERT approach will be replicated using the precedence method of scheduling and demonstrated on an example.

13.2 Network construction

In its traditional format, PERT is concerned with specific events or milestones that are important to accomplish and against which progress is measured. When these events or milestones have been defined, they are linked together by arrows to form a PERT network, which in fact is identical to the arrow network that was briefly discussed in Chapter 3.

While PERT and the arrow method of CPM share the same network, information they generate is interpreted differently. The arrow method of CPM focuses on activities that are to be accomplished while PERT is concerned with events. For example, events 1–2 in Figure 13.1 define the activity 'Excavate site' in CPM, while in PERT event 1 may relate to 'Contract awarded' or 'Start excavation' and event 2 to 'Site excavated'. The difference in interpreting PERT and CPM is illustrated in Table 13.1.

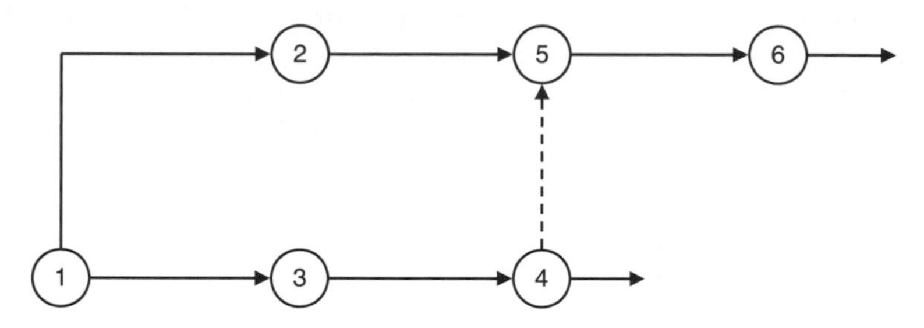

Figure 13.1 A PERT network.

Table 13.1 A comparison of interpretation of the arrow method of CPM and the PERT method

CPM activity	*Arrow method activity description*
1–2	Excavate site
1–3	Excavate for sewerage and drainage services
3–4	Install sewerage and drainage services
2–5	Form slab-edge beams
5–6	Concrete to ground floor slab
PERT event	*PERT event description*
1	Contract awarded, start excavation
2	Site excavated, formwork started
3	Services excavated, installation started
4	Services installed
5	Slab-edge beams formed, services installed, concrete to slab started
6	Ground floor slab concreted

13.3 The probability concept in PERT

The fundamental modelling and computational difference between PERT and CPM is related to the expression of duration of activities (it should be remembered that in PERT, an activity is treated as a pair of events). CPM defines duration of activities deterministically as a single-value estimate, while PERT, being probabilistic in nature, assumes that activity duration is a random variable with relatively large variances. In PERT, distribution of activity duration is assumed to fall on a beta probability distribution curve. Sasieni (1986) and others have questioned the validity of this assumption, but no firm consensus has yet emerged on the choice of a more appropriate distribution for modelling durations of activities in PERT.

Beta distribution is described by its mean and standard deviation. These two parameters are calculated from three estimates of activity duration, where:

m = the most likely time interval between two events but not necessarily the mean.
a = the most optimistic time. This is the shortest time interval and is equivalent to a chance of 1 in 100 that the time estimate will be achieved.
b = the most pessimistic time estimate. This is the longest time interval representing the worst scenario and is equivalent to a chance of 1 in 100 that the estimate will be this bad.

These three time estimates may be derived from a database or, as is commonly the case, subjectively by 'experts'. The mean 't_e' of beta distribution is expressed in the following formula:

$$t_e = \frac{a + 4m + b}{6}$$

The standard deviation 's' of beta distribution is expressed as follows:

$$s = \frac{b - a}{6}$$

In finding the expected duration of the project, PERT relies on a widely used statistical concept known as the Central Limit Theorem. Harris (1978: 326) describes the theorem in the following terms:

> If independent probability distributions are to be summed, then the mean of the sum is the sum of the individual means, the variance of the sum is the sum of the individual variances, and the distribution of the sum tends to the shape of the normal curve regardless of the shape of the individual input distributions.

According to this theorem, the expected project duration 'T_e' is a sum of mean 't_e' durations of critical activities in a schedule, provided the number of critical activities is reasonably large.

$$T_e = \Sigma(t_e) \text{ for critical activities}$$

Similarly, the standard deviation 'S' of the distribution of the expected project duration is according to the theorem derived as a sum of a square root of squared standard deviations (variances) of individual critical activities.

$$S = \sqrt{\Sigma s^2} \text{ for critical activities}$$

The values 'T_e' and 'S' are the mean and the standard deviation of the normal distribution by which the normal distribution is fully defined. It means that for a different set of mean and standard deviation values there is a unique normal curve. Another interesting characteristic of the normal distribution is its bell-like shape, which is smooth and symmetrical around its mean. Although the curve extends indefinitely in either direction from the mean, 99.7 per cent of the area under the curve lies between −3 and +3 standard deviations from the mean. Within two standard deviations either way from the mean lies 95.5 per cent of the area, and 68 per cent of the area lies within one standard deviation either way from the mean. The steeper the curve, the smaller the dispersion and, conversely, the flatter the curve, the greater the dispersion of values from the mean. These areas under a normal curve represent the probability of possible outcomes. The interpretation is that, for example, the project duration that falls within one standard deviation either way from the mean has a 68 per cent probability of being achieved.

The characteristics of a normal curve permit calculation of probabilities of values that are not exactly one, two or three standard deviations from the mean. The first step is to establish a 'z' score, which is the distance of a particular value from the mean. It is calculated as follows from the following formula:

$$z = \frac{x - \mu}{\sigma}$$

where:

z = the distance from the mean (in standard deviations)
x = a particular value in question
μ = the mean of a normal distribution
σ = the standard deviation of a normal distribution.

A 'z' value is then converted to a probability value using a probability table, given in Table 13.2. For example, a z score of +1.36 is the distance 1.36 standard deviations to the right of the mean.

Table 13.2 The probability table for a normal distribution

z	*0*	*1*	*2*	*3*	*4*	*5*	*6*	*7*	*8*	*9*
0.0	0.5000	0.5040	0.5080	0.5120	0.5160	0.5199	0.5239	0.5279	0.5319	0.5359
0.1	0.5398	0.5438	0.5478	0.5517	0.5557	0.5596	0.5636	0.5675	0.5714	0.5754
0.2	0.5793	0.5832	0.5871	0.5910	0.5948	0.5987	0.6026	0.6064	0.6103	0.6141
0.3	0.6179	0.6217	0.6255	0.6293	0.6331	0.6368	0.6406	0.6443	0.6480	0.6517
0.4	0.6554	0.6591	0.6628	0.6664	0.6700	0.6736	0.6772	0.6808	0.6844	0.6879
0.5	0.6915	0.6950	0.6985	0.7019	0.7054	0.7088	0.7123	0.7157	0.7190	0.7224
0.6	0.7258	0.7291	0.7324	0.7356	0.7389	0.7422	0.7457	0.7486	0.7518	0.7549
0.7	0.7580	0.7612	0.7642	0.7673	0.7704	0.7734	0.7764	0.7794	0.7823	0.7852
0.8	0.7881	0.7910	0.7939	0.7967	0.7996	0.8023	0.8051	0.8078	0.8106	0.8133
0.9	0.8159	0.8186	0.8212	0.8238	0.8264	0.8289	0.8315	0.8340	0.8365	0.8389
1.0	0.8413	0.8438	0.8461	0.8485	0.8508	0.8531	0.8554	0.8577	0.8599	0.8621
1.1	0.8643	0.8665	0.8686	0.8708	0.8729	0.8749	0.8770	0.8790	0.8810	0.8830
1.2	0.8849	0.8869	0.8888	0.8906	0.8925	0.8944	0.8962	0.8980	0.8997	0.9015
1.3	0.9032	0.9049	0.9066	0.9082	0.9099	0.9115	0.9131	0.9147	0.9162	0.9177
1.4	0.9192	0.9207	0.9222	0.9236	0.9251	0.9265	0.9279	0.9292	0.9306	0.9319
1.5	0.9332	0.9345	0.9357	0.9370	0.9382	0.9394	0.9406	0.9418	0.9430	0.9441
1.6	0.9452	0.9463	0.9474	0.9484	0.9495	0.9505	0.9515	0.9525	0.9535	0.9545
1.7	0.9554	0.9564	0.9573	0.9582	0.9591	0.9599	0.9608	0.9616	0.9625	0.9633
1.8	0.9641	0.9649	0.9656	0.9664	0.9671	0.9678	0.9686	0.9693	0.9699	0.9706
1.9	0.9713	0.9719	0.9726	0.9732	0.9738	0.9744	0.9750	0.9756	0.9761	0.9767
2.0	0.9772	0.9778	0.9783	0.9788	0.9793	0.9798	0.9803	0.9808	0.9812	0.9817
2.1	0.9821	0.9826	0.9830	0.9834	0.9838	0.9824	0.9846	0.9850	0.9854	0.9857
2.2	0.9861	0.9864	0.9868	0.9871	0.9875	0.9878	0.9881	0.9884	0.9887	0.9890
2.3	0.9893	0.9896	0.9898	0.9901	0.9904	0.9906	0.9909	0.9911	0.9913	0.9916
2.4	0.9918	0.9920	0.9922	0.9925	0.9927	0.9929	0.9931	0.9932	0.9934	0.9936
2.5	0.9938	0.9940	0.9941	0.9943	0.9945	0.9946	0.9948	0.9949	0.9951	0.9952
2.6	0.9953	0.9955	0.9956	0.9957	0.9959	0.9960	0.9961	0.9962	0.9963	0.9964
2.7	0.9965	0.9966	0.9967	0.9968	0.9969	0.9970	0.9971	0.9972	0.9973	0.9974
2.8	0.9974	0.9975	0.9976	0.9977	0.9977	0.9978	0.9979	0.9979	0.9980	0.9981
2.9	0.9981	0.9982	0.9982	0.9983	0.9984	0.9984	0.9985	0.9985	0.9986	0.9986
3.0	0.9987	0.9987	0.9987	0.9988	0.9988	0.9989	0.9989	0.9989	0.9990	0.9990
3.1	0.9990	0.9991	0.9991	0.9991	0.9992	0.9992	0.9992	0.9992	0.9993	0.9993
3.2	0.9993	0.9993	0.9994	0.9994	0.9994	0.9994	0.9994	0.9995	0.9995	0.9995
3.3	0.9995	0.9995	0.9996	0.9996	0.9996	0.9996	0.9996	0.9996	0.9996	0.9997
3.4	0.9997	0.9997	0.9997	0.9997	0.9997	0.9997	0.9997	0.9997	0.9998	0.9998
3.5	0.9998	0.9998	0.9998	0.9998	0.9998	0.9998	0.9998	0.9998	0.9998	0.9998
3.6	0.9998	0.9998	0.9999	0.9999	0.9999	0.9999	0.9999	0.9999	0.9999	0.9999

Apart from the criticism that a beta distribution may not accurately model a distribution of values of activity durations, others, notably Keefer and Bodily (1983) and Sasieni (1986), have questioned the correctness of the PERT method, particularly with regard to the application of the Central Limit Theorem and the assumption of independence. Another problem alluded to by Sculli (1983), Kuklan *et al.* (1993), Gong and Hugsted (1993) and Ranasinghe (1994) concerns an optimistically biased estimation of project time. This is because PERT assumes that the critical path cannot change for as long as other non-critical paths have float. However, it may well be that risk associated with non-critical events/activities is greater than that of critical activities. Should this risk eventuate, the available float may be insufficient to absorb any ensuing delays. In PERT, the more float there is on parallel non-critical paths, the better the probability of accomplishing the schedule on time.

13.4 The PERT method modelled on the precedence method of CPM

The original concept of PERT, based on the arrow CPM method, is event-oriented. Its main aim is to ensure completion of events or milestones on time. In CPM scheduling, the arrow method has been superseded by the precedence method, which is not only easier to work with but is also better supported by quality computer software. Given that the precedence method has become the preferred method of CPM scheduling, its suitability for use with the PERT method is worth investigating.

The most obvious difference between the arrow and precedence methods of CPM scheduling is the lack of events in the precedence method. It means that rather than being concerned with events, a precedence-based PERT would focus on activities. This is probably of little relevance to the final outcome provided the planner is able to adapt to the change.

In PERT, float is referred to as slack. The original concept of PERT expresses two different types of slack: activity slack and event slack. In a precedence-based PERT it would be possible to express free activity slack as free and total slack, which would be the same as free and total float, but event slack would not be applicable.

PERT adopts the computational process defined in Chapter 3. After a PERT schedule is created, the planner determines activity durations. These are assigned to activities as mean t_e values. The planner then performs the forward and backward pass calculations, which determine the position of a critical path, calculates the activity slack and calculates the probability of the schedule being completed on time.

Let's test the applicability of the precedence method in PERT on the project illustrated in Figure 13.2. The contractor is required by the contract to complete refurbishment of the project in 22 weeks (T_s = 22 weeks). The client wants to know the probability that the project will be completed on time.

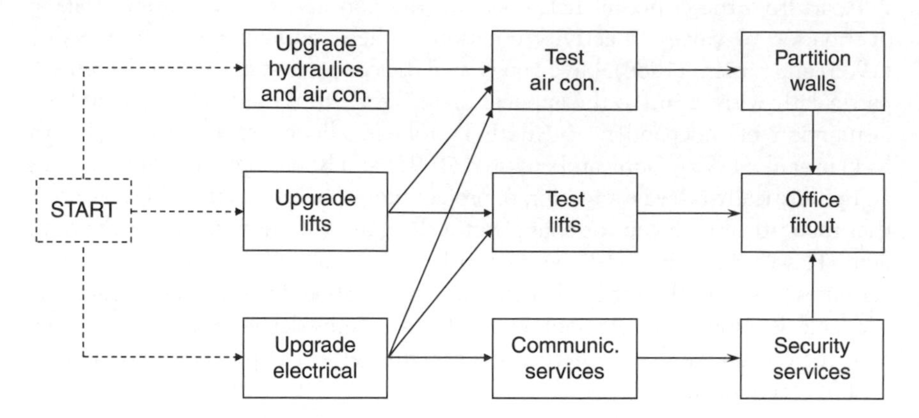

Figure 13.2 A precedence-based PERT schedule of the refurbishment project.

Table 13.3 Three time estimates, a, m and b, of activity durations and t_e, s and s^2 values

Activity	*a*	*m*	*b*	t_e	s	s^2
Upgrade electrical	5	6	9	6.33	0.67	0.44
Upgrade lifts	8	9	12	9.33	0.67	0.44
Upgrade hydraulics and air con.	4	5	7	5.17	0.50	0.25
Communication services	4	5	8	5.33	0.67	0.44
Test lifts	2.5	3	4	3.08	0.25	0.06
Test air con.	1.5	2	5	2.42	0.58	0.34
Partition walls	3	4	6	4.17	0.50	0.25
Security services	2.5	3	4	3.08	0.25	0.06
Office fitout	2	3	5	3.17	0.50	0.25

Estimates of durations of activities together with the values of mean t_e, standard deviation s, and variance s^2 for each activity are given in Table 13.3.

13.4.1 Step 1

From the three time estimates of activity durations, the planner calculates the mean, standard deviation and variance values of time distributions of each activity in the schedule. These values are given in Table 13.3. In PERT, the mean values t_e represent durations of activities.

13.4.2 Step 2

The planner performs the forward and the backward pass calculations of the schedule, determines the critical path, and calculates values of activity slack. The overall project duration T_e becomes the mean value of the output normal distribution.

The calculated schedule is illustrated in Figure 13.3. The expected completion time of the project T_e = 19.09 weeks. Slack values of non-critical activities are calculated as free and total float.

13.4.3 Step 3

According to the Central Limit Theorem, the completion time of the project is normally distributed with the mean T_e and the standard deviation S. These two values are now calculated as follows:

T_e = 19.09 weeks

$S = \sqrt{\Sigma s^2}$ for the critical activities (Upgrade lifts; Test air con.; Partition walls; Office fitout)

$S = \sqrt{(0.44 + 0 + 0.34 + 0.25 + 0.25)}$

$S = \sqrt{1.28}$

$S = 1.13$

The probability z of completing the project in 22 weeks is calculated from the following formula:

$$z = \frac{x - \mu}{\sigma}$$

where:

$x = T_s$
$\mu = T_e$
$\sigma = S$

$$z = (22 - 19.09)/1.13 = 2.58$$

In the probability table, a z score of 2.58 is read off as 99.5 per cent probability. Therefore, there is an almost 100 per cent probability that the project will be completed in 22 weeks.

If, for example, the scheduled completion date of the project is only 20 weeks, a z score is:

$$z = (20 - 19.09)/1.13 = 0.81$$

The project would then have just under 80 per cent probability of being completed in 20 weeks.

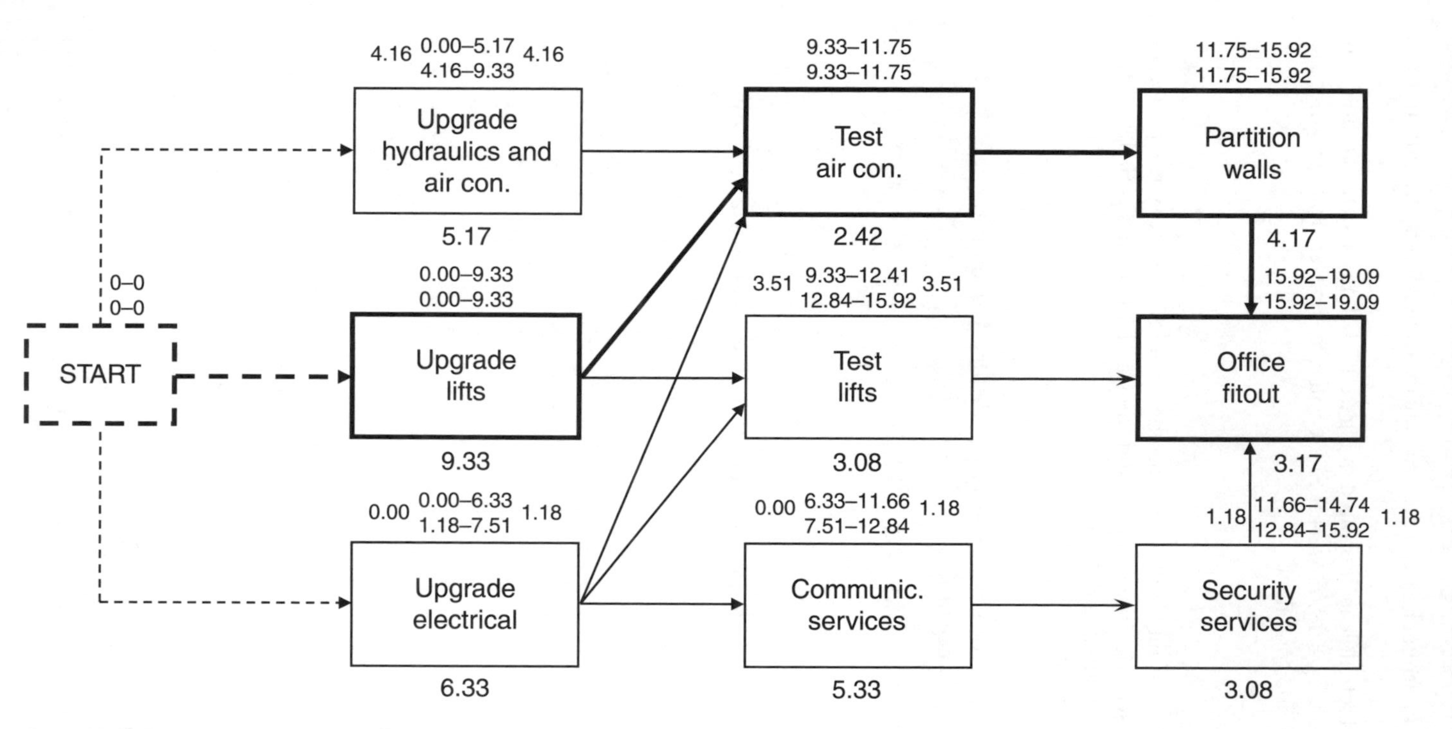

Figure 13.3 The calculated PERT schedule.

13.5 Summary

This section has introduced the concept of probability scheduling and control using PERT. Although the original concept of PERT was developed around the arrow method of CPM to track progress of events or milestones, the PERT method works equally well when based on the precedence method of CPM.

The PERT method, particularly the use of beta distribution and the assumption of independence among activities in a schedule, has frequently been criticised. Nevertheless, PERT's value in planning and evaluating large, particularly military, projects has successfully been demonstrated in the past.

Exercises

Solutions to the following exercises can be found on the following website: http://www.routledge.com/books/details/9780415601696/

Exercise 13.1

Standard deviations and variances of activities in a project are given in Table 13.4. What is the probability (z) that the project will be completed within 25 weeks? The project duration T_e has been calculated as 23.5 weeks.

Exercise 13.2

A PERT schedule (Figure 13.4) shows a sequence of activities related to the construction of a small building project. Information on durations of activities and values of means, standard deviations and variances are given in Table 13.5. What is the probability that this project will be completed in 31 weeks?

Table 13.4 Data for the PERT example

Activity	s	s^2
A	1.00	1.00
B	**1.33**	**1.77**
C	0.67	0.45
D	**0.83**	**0.69**
E	**0.50**	**0.25**
F	0.50	0.25
G	**0.83**	**0.69**
H	1.50	2.25

Note: Activities B, D, E and G are critical.

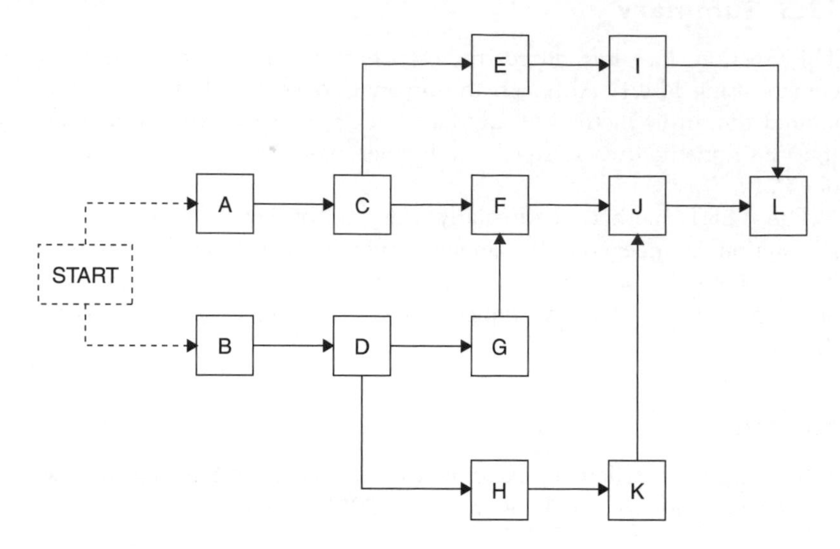

Figure 13.4 The PERT schedule.

Table 13.5 Data for the PERT example

Activity	*a*	*m*	*b*	t_e	*s*	s^2
A	6	9	15	9.5	1.50	2.25
B	2	4	8	4.3	1.00	1.00
C	6	8	10	8.0	0.67	0.45
D	4	7	12	7.3	1.33	1.77
E	2	3	6	3.3	0.67	0.45
F	1	2	4	2.2	0.50	0.25
G	4	7	9	6.8	0.83	0.69
H	5	9	11	8.7	1.00	1.00
I	1	4	6	3.8	0.83	0.69
J	2	4	5	3.8	0.50	0.25
K	2	3	5	3.2	0.50	0.25
L	2	3	5	3.2	0.50	0.25

Exercise 13.3

Information on durations of activities in a PERT schedule (Figure 13.5), and values of means, standard deviations and variances are given in Table 13.6. What is the probability that this project will be completed in 32 weeks?

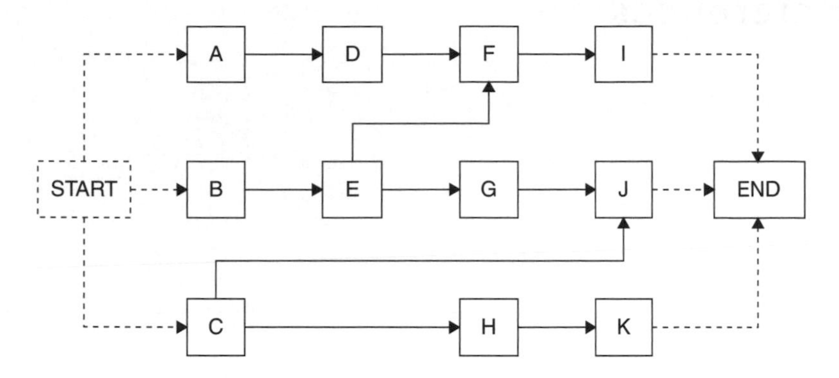

Figure 13.5 The PERT schedule

Table 13.6 Data for the PERT example

Activity	*a*	*m*	*b*	t_e	*s*	s^2
A	2	3	5	3.2	0.50	0.25
B	6	8	14	8.7	1.33	1.78
C	4	7	11	7.2	1.17	1.36
D	5	7	8	6.8	0.50	0.25
E	4	6	10	6.3	1.00	1.00
F	7	10	12	9.8	0.83	0.69
G	6	9	13	9.2	1.17	1.36
H	3	5	9	5.3	1.00	1.00
I	2	5	7	4.8	0.83	0.69
J	3	5	8	5.2	0.83	0.69
K	8	12	17	12.2	1.50	2.25

References

Al-Harbi, K. A., Selim, S. Z. and Al-Sinan, M. (1996) 'A multiobjective linear program for scheduling repetitive projects', *Cost Engineer*, 38(12) December: 41–45.

Anon (2002) 'Critical chain basics', *Focused Performance*, *http://focusedperformance.com/articles/cc01.html* 1–3.

AS (1992) *Handbook symbols and abbreviations for building and construction*, Standards Australia HB 24–1992.

AS/NZS (2004) *Australian/New Zealand standard on risk management*, Standards Australia and Standards New Zealand AS/NZS 4360:2004.

Betts, M. (1991) 'Achieving and measuring flexibility in project information retrieval', *Construction Management and Economics*, 9(3): 231–245.

BS (1992) *Glossary of terms used in management services BS 3138*, British Standards Institution, London.

Bryson, J. M. and Alston, F. K. (2004) *Creating and implementing your strategic plan: a workbook for public and nonprofit organisations*, John Wiley & Sons, New York.

Burke, R. (1999) *Project management: planning and control techniques*, 3rd ed., John Wiley & Sons, Chichester.

Byrne, P. and Cadman, D. (1984) *Risk, uncertainty and decision-making in property development*, E & FN Spon, London.

Chitkara, K. K. (1998) *Construction project management – planning, scheduling and controlling*, Tata McGraw-Hill, New Delhi.

Christensen, D. S. (2011) 'Earned value bibliography', *http://www.suu.edu/faculty/ChristensenD/EV-bib.html*

Cooper, D. and Chapman, C. (1987) *Risk analysis for large projects*, John Wiley & Sons, London.

Currie, R. M. (1959) *Work study*, Pitman, New York.

Dawson, P. and Palmer, G. (1995) *Quality management – the theory and practice of implementing change*, Longman, Melbourne.

Edwards, L. (1995) *Practical risk management in the construction industry*, Thomas Telford, London.

Flanagan, R. and Norman, G. (1993) *Risk management and construction*, Blackwell Science, Oxford.

Gilmour, P. and Hunt, R. A. (1995) *Total quality management: integrating quality into design, operations and strategy*, Longman, Melbourne.

Goedert, J. D. and Meadati, P. (2008) 'Integrating construction process documentation into building information modeling', *Journal of Construction Engineering and Management*, ASCE, 134(7): 509–516.

Goldratt, E. M. (1997) *Critical chain*, North River Press, Great Barrington, MA, USA.

Gong, D. and Hugsted, R. (1993) 'Time-uncertainty analysis in project networks with a new merge-event time estimation technique', *International Journal of Project Management*, 11(3): 165–174.

Hamilton, A. (1997) *Management by projects*, Thomas Telford, London.

Harris, F. and McCaffer, R. (1991) *Management of construction equipment*, 2nd ed., MacMillan, London.

Harris, F. and McCaffer, R. (2006) *Modern construction management*, 6th ed., Blackwell Science, Oxford.

Harris, R. (1978) *Precedence and arrow networking techniques for construction*, John Wiley & Sons, New York.

Healy, P. L. (1999) *Project management: getting the job done on time and in budget*, Butterworth-Heinemann, Victoria, Australia.

Hertz, D. B. and Thomas, H. (1983) 'Decision and risk analysis in a new product and facilities planning problem', *Sloan Management Review*, Winter: 17–31.

Horowitz, J. (1967) *Critical path scheduling*, Ronald Press Company, New York.

Ivkovic, B. (1991) 'Project modelling and uncertainty', *Proceedings of International Conference on Construction Project Modelling and Productivity*, Dubrovnik, 15–32.

Keefer, D. L. and Bodily, S. E. (1983) 'Three-point approximations for continuous random variables', *Management Science*, 29(5): 595–609.

Kuklan, H., Erdem, E., Nasri, F. and Paknejad, M. J. (1993) 'Project planning and control: an enhanced PERT network', *International Journal of Project Management*, 11(2): 87–92.

Lewis, J. (2001) *Project planning, scheduling and control*, McGraw-Hill, USA.

Loosemore, M., Raftery, J., Reilly, C. and Higgon, D. (2006) *Risk management in projects*, 2nd ed., Taylor & Francis, Abingdon, UK.

Malcomb, D. G., Roseboom, J. H., Clark, C. E. and Fazar, W. (1959) 'Applications of a technique for research and development program evaluation (PERT)', *Operations Research*, 7: 646–649.

Mason, G. E. (1973) *A quantitative risk management approach to the selection of construction contract provisions*, Technical Report No. 173, The Construction Institute, Department of Civil Engineering, Stanford University, Palo Alto, CA, USA.

Mattila, K. G. and Abraham, D. M. (1998) 'Linear scheduling: past research efforts and future directions', *Engineering, Construction and Architectural Management*, 5(3): 294–303.

McGeorge, D. and Palmer, A. (2002) *Construction management – new directions*, 2nd ed., Blackwell Science, Oxford.

Newbold, R. C. (1998) *Project management in the fast lane: applying the theory of constraints*, St. Lucie Press, Boca Raton, FL, USA.

NSW Government (2004) *NSW Government procurement policy*, NSW Treasury, Sydney.

NSW Government (2009) *Internal audit and risk management policy for the NSW public sector*, NSW Government, Sydney.

Oakland, J. S. and Porter, L. (1994) *Cases in total quality management*, Butterworth-Heinemann, Oxford.

Oakland, J. S. and Sohal, A. S. (1996) *Total quality management – text with cases*, Butterworth-Heinemann, Melbourne.

O'Brien, J. J. (1975) 'CPM scheduling for high-rise buildings', *Journal of the Construction Division*, ASCE, 101(4): 895–908.

Oracle Corporation (2009) *Primavera version 6.2.1 – Primavera P6 project management reference manual*, Oracle Corporation, Sydney.

Oxley, R. and Poskitt, J. (1980) *Management techniques applied to the construction industry*, Crosby Lockwood, London.

Patrick, F. S. (1999) 'Critical chain scheduling and buffer management: getting out from between Parkinson's rock and Murphy's hard place', *http://focusedperformance.com/articles/ccpm.html*

Pfeffer, J. (1998) *The human equation: building profits by putting people first*, Harvard Business School Press, Boston.

Porter, C. E. (1981) *Risk allocation in construction contracts*, MSc Thesis, University of Manchester, Manchester, UK.

Raftery, J. (1990) 'Risk analysis for construction forecasting', *Proceedings of Risk Analysis for Construction Forecasting Workshop*, Department of Surveying, Thames Polytechnic, London.

Ranasinghe, M. (1994) 'Quantification and management of uncertainty in activity duration networks', *Construction Management and Economics*, 12(1):15–29.

Robbins, S. and Coulter, M. (2009) *Management*, 10th ed., Prentice Hall, Sydney.

Sasieni, M. W. (1986) 'A note on PERT times', *Management Science*, 32(12): 1652–1653.

Sculli, D. (1983) 'The completion time of PERT networks', *Journal of Operational Research Society*, 34(2): 155–158.

Spence, J. (1980) 'Modern risk management concepts', *BSFA Conference on Risk Management in Building*, Sydney, 21–29.

Townley, T. W. (1991) 'Historical data-driven range estimating', *Transactions of the 1991 American Association of Cost Engineers*, 35th Annual Meeting: G.2.1–G.2.5, Seattle, WA, USA.

Uher, T. E. and Davenport, P. (2009) *Fundamentals of building contract management*, 2nd ed., UNSW Press, Sydney.

Wilson, A. (1984) 'Risk analysis in retail store design: a case study', *Transactions of Design Quality Cost Profit Conference*, Portsmouth Polytechnic, Portsmouth, UK.

Index